GREEN CARBON

THE ROLE OF NATURAL FORESTS IN CARBON STORAGE

Part 2. Biomass carbon stocks in the Great Western Woodlands

Sandra Berry, Heather Keith, Brendan Mackey,
Matthew Brookhouse, and Justin Jonson*

The Fenner School of Environment & Society, The Australian National University

* Greening Australia, Western Australia

ANU
THE AUSTRALIAN NATIONAL UNIVERSITY

E PRESS

ANU
E PRESS

Published by ANU E Press
The Australian National University
Canberra ACT 0200, Australia
Email: anuepress@anu.edu.au
Web: http://epress.anu.edu.au
Online version available at: http://epress.anu.edu.au/green_carbon2_citation.html

National Library of Australia Cataloguing-in-Publication entry

Title: Green carbon : the role of natural forests in carbon
            storage Part 2. Biomass carbon stocks in the Great
            Western Woodlands / Sandra. Berry ... [et al.]

ISBN: 9781921666704 (pbk) (Part 2)
            9781921666711 (online)

Subjects: Carbon--Environmental aspects.
            Forests and forestry--Environmental aspects.
            Plants--Effect of atmospheric carbon dioxide on.
            Carbon dioxide mitigation.

Other Authors/Contributors: Berry, Sandra L.

Dewey Number: 577.3144

Printed by Griffin Press

This edition © 2010 ANU E Press

Design by ANU E Press

Cover photograph: Eucalyptus woodland, Great Western Woodlands, Western Australia, Sandra Berry

CONTENTS

ACKNOWLEDGMENTS

This book is based on a report supported by a grant from The Wilderness Society. We would like to thank ANU Enterprise Pty Ltd, which has been involved in the management of the project and has provided support that has been invaluable to the team. Significant components of this report, in particular the field-based plant measurements and the F_{PAR} time-series data-set, drew on work funded by Australian Research Council Linkage Project LP0455163 for which The Wilderness Society was an industry partner. Greening Australia kindly made available recently derived equations for some woodland trees and mallee in south-western Western Australia. Professor Ross Bradstock and Dr Rod Fensham provided peer reviews of a draft of this report. We thank them for their constructive comments, which have led to a more comprehensive final document. We thank two anonymous reviewers for their comments on the final document. We thank Ian Smith for his assistance with research into the history of timber cutting in the Great Western Woodlands.

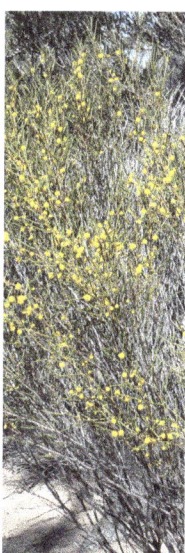

EXECUTIVE SUMMARY

1. The 'Great Western Woodlands' (GWW) includes most of the contiguous residual natural woody vegetation to the east of the wheatbelt in south-western Western Australia. The conservation status and future of the region are being assessed by various government and non-government stakeholders, and there is a growing need for more quantitative understanding of the environmental services provided by GWW ecosystems—in particular, their role in carbon storage. Our results show that the natural vegetation of woodland and shrubland has been extensively modified by: 1) changed fire regimes (mostly human-induced); 2) timber cutting; and 3) mining and mineral exploration. The condition of GWW natural vegetation and prospects for its ecological restoration need to be factored into conservation planning and related policies and measures for natural resource management.

2. Biomass carbon stocks were calculated using measurements of tree dimensions from 21 field sites covering different vegetation communities within the GWW. Because there is no information specifically for the GWW, allometric equations that relate tree dimensions to biomass were compared from several studies of woodlands in Australia. Equations for woodland sites in Queensland (Burrows et al. 2000, 2002), Northern Territory (Williams et al. 2005) and the wheatbelt in Western Australia (Jonson unpublished) were statistically sufficiently similar to be considered equivalent. The equation from Williams et al. (2005) was selected to convert the GWW site data into estimates of biomass because it included the variables of height and diameter, which account for variations in tree form that are likely due to variations in environmental conditions and as the result of tree damage.

3. Biomass estimated at field sites was extrapolated across the region using: 1) spatial data layers of mapped vegetation types (DEWHA 2005); 2) the fraction of photosynthetically active radiation intercepted by the green foliage of the evergreen vegetation canopy (F_E) derived from MODIS satellite imagery (Paget and King 2008); and 3) disturbance history information about land-use activities associated with timber cutting and mining. Biomass at the 21 field sites was related to F_E within four categories of land use and vegetation type: 1) no mineral exploration and no timber cutting in eucalypt woodlands; 2) mineral exploration and no timber cutting in eucalypt woodlands; 3) timber cutting in eucalypt woodlands with or without mineral exploration; and 4) shrublands and non-eucalypt woodlands. Biomass was calculated from the value of F_E in each pixel within the spatial extent of each of these four categories. Average total biomass (above and below-ground, living and dead biomass) was 21 t C ha⁻¹ and the total for the GWW region was 312 Mt C[1].

4. We mapped the footprints of extremely severe fires in the GWW over the past 35 years. Wildfire footprints were detected through the analysis of time-series of satellite imagery (1972–2007) and areas that had been burnt once, twice and three times during this period were mapped. These fires have resulted in the death of mature eucalypt tree stems over extensive areas and have initiated a pyric successional vegetation of seedlings and re-sprouts from roots. The pre-fire eucalypt woodland[2] vegetation structure has, following fire, changed to a pyric successional eucalypt or acacia-dominated shrubland[3] vegetation structure.

---

1   M (mega) is one million, or in scientific notation, 1×10^6.
2   Woodland: height of tallest stratum >10 m; low woodland: height of tallest stratum 5–10 m.
3   Shrubland: height of tallest stratum <5 m.

5. We conclude from our analysis that current vegetation types comprise mostly natural eucalypt woodland and pyric successional stages of shrubland dominated by eucalypt or acacia[4]. These stages are demonstrated by the congruence between boundaries of vegetation mapping units and fire footprints. In accordance with this proposition, we have modified the mapped distribution of eucalypt woodland to account for the changes due to fires that post-date vegetation mapping. This modification involved reclassifying the 1.46 million hectares of eucalypt woodland that had been severely burnt by fires between 1972 and 2007 into tall closed shrublands—a vegetation class that better describes the present pyric successional vegetation structure of eucalypt seedlings, saplings and lignotuber re-sprouts (mallee).

6. Disturbances by human land-use activities that have reduced biomass carbon stocks in the GWW include: 1) increased incidence of fire; 2) mining and mineral exploration on the greenstone lithology; 3) timber cutting; and 4) pastoralism. Field measurement-based estimates indicate that the current above-ground biomass of woodlands recovering from timber cutting (mostly for bio-fuel) in the first half of the past century could be at just 40–50 per cent of the above-ground biomass at carbon carrying capacity. Similarly, the above-ground biomass carbon of woodlands impacted by mineral exploration could be at about 70 per cent of the value at carbon carrying capacity. The change in fire regimes is the major factor. Fire occurrence has increased due to human land-use activities within the GWW resulting in increases in ignition sources. Regeneration of eucalypts after fire as seedlings or re-sprouts produces denser vegetation, providing a more continuous fuel layer that carries more intense fires. These disturbance factors have, over extensive areas, both reduced carbon stocks within vegetation types and changed vegetation structure from woodland to pyric successional stages of dense mallee, marlock[5] and acacia-dominated shrubland. In a hypothetical scenario of 'no-disturbance' condition—if the woodlands had not been impacted on by fire, timber cutting, mineral exploration and pastoral land management—where eucalypt woodland occupied all the area currently assumed to be under pyric successional stages, a total of 13 million ha (double the current extent and 80 per cent of the GWW) would be woodland, and the total biomass carbon stock would be 915 Mt C (triple the current stock).

7. Soil carbon is the largest pool of terrestrial carbon and some components of the soil carbon store have great longevity. Few data exist to quantify soil carbon in woodlands and specifically in the GWW. We used the spatial data sets of the Australian Soil Resource Information System (CSIRO 2007) and field soil data from Wynn et al. (2006) to estimate an average soil carbon stock for the GWW of 40 t C ha^{-1} and total carbon stock for the A and B horizons of 639 Mt C. Soil carbon stocks likely have been reduced in areas where disturbance has reduced biomass. These estimates of soil carbon can, however, be considered only as indicative; further field survey is a priority.

---

4   Note that heath and other non-eucalypt dominated shrublands and woodlands cover approximately 1.2 million ha of the GWW. Although we have estimated the current carbon stock of these communities, we have not developed a model to estimate their carbon carrying capacity under a no-disturbance scenario as we have no field data for the undisturbed or 'climax' state of these vegetation types. For these vegetation types, the carbon carrying capacity is equated to the current carbon stock.

5   Marlock is the term used to describe dwarf eucalypts having a single stem or mallee-like form but with poor development of a lignotuber.

8. The estimated total carbon stock of the soil and vegetation in the GWW as of January 2008 was 950 Mt C. Under the hypothetical 'no-disturbance' condition, the total carbon stock is estimated at 1550 Mt C.

9. Maximising carbon stocks in the GWW depends on both avoiding emissions from further degradation and restoring currently degraded eucalypt woodlands. The estimated total biomass carbon carrying capacity of 915 Mt C within the GWW (for the 'no-disturbance' scenario) should be considered as a maximum possible store of biomass carbon if intense fire had been excluded for several centuries and there had been no timber cutting, mineral exploration or other anthropogenic causes of woodland thinning. Given that the age of trees in undisturbed woodland sites is estimated to cover a range from 100 to 400 years (plus some trees that are very large and possibly very ancient), re-growth of pyric successional stages of vegetation to attain the structure and biomass of eucalypt woodlands would take a very long time.

10. The management of the GWW to maximise carbon stocks would require a substantial reduction in fire frequency and intensity. Ideally, the entire region should be placed under a conservation management planning overlay— perhaps analogous to the management of the Great Barrier Reef in its entirety. Changing land tenure from vacant crown land to nature conservation reserve would provide for jurisdiction by the WA Department of Environment and Conservation (WADEC). Management activities by WADEC would need to include restricting vehicular access, providing additional resources for early detection and suppression of fires, imposition of total fire ban periods during conditions of high fire danger and establishing an education campaign to reduce accidental and deliberate ignition events.

11. Conservation management options are most likely to succeed if they can be linked to the emerging carbon market and payments for land stewardship and ecosystem services. Therefore, it is important that state, Commonwealth and international policies and actions recognise the value of avoiding emissions from extant carbon stocks in the GWW, along with the sequestration potential from managing threatening processes. Incentives are needed that will enhance carbon stocks in the GWW through ecological restoration, while avoiding perverse outcomes such as inadvertently providing incentives to clear and degrade natural vegetation ecosystems.

1. INTRODUCTION

1.1 BACKGROUND

The region referred to as the 'Great Western Woodlands' (hereafter, GWW) includes most of the contiguous residual natural woody vegetation to the east of the wheatbelt in south-western Western Australia. Before European settlement, south-western Western Australia supported a natural woody vegetation cover of woodland and shrubland (Figure 1.1a). In the past two centuries, about half of this woody vegetation has been cleared and replaced with agricultural production—mostly wheat (Figure 1.1b)—and is commonly referred to as the 'wheatbelt'. The GWW region to the east of the wheatbelt was found to be less suitable for agricultural crops or livestock grazing. Nonetheless, development of the GWW continues to be proposed. In 1979, the WA Government proposed an extension of farming across the southern half of the GWW (Bradby 2008). In 1992, the WA Department of Conservation and Land Management and Goldfields residents established the Goldfields Specialty Timber Industry Group Incorporated with the aim of exploiting the timber resources of the GWW (Siemon and Kealley 1999). Despite early attempts at development, the vegetation remains largely in a natural state in that it has not been subject to broad-scale land clearing and intensive agricultural development. The region, however, overlies the Late Archaean granite-greenstone terranes of the Eastern Yilgarn Craton—geological formations containing high-quality gold and nickel deposits. Consequently, much of the region has been impacted on by mineral exploration and mining. Major impacts on the natural vegetation include: 1) timber cutting before 1975 to provide mining timber and firewood for the goldfields and water-pumping stations; 2) the creation of numerous access tracks and seismic lines; and 3) changes to the fire regime. The impact of changed fire regimes demands special attention as previous studies show that this can lead to the conversion of woodland to a mallee–shrubland vegetation type (Hopkins and Robinson 1981).

The current and future conservation status of the GWW is of increasing concern and there is a need to gain a more quantitative understanding of the environmental services the region provides. The biodiversity significance of the GWW, including its high levels of endemic plant species and the importance of large areas as bird habitat, is known (Duncan et al. 2006; Newby et al. 1984; Recher et al. 2007). Particularly important, however, in the context of the climate change problem, is the role of the GWW's vegetation ecosystems in carbon storage. Carbon storage in natural ecosystems of the biosphere is referred to as 'green carbon' (Mackey et al. 2008). Green carbon is stored in woody plant tissues of living trees and shrubs, dead stems, coarse woody debris on the ground and organic matter (derived from decomposition of plant tissues) in the upper soil layers. The GWW is a large area of natural woodland that potentially includes a significant component of the green carbon stock within Western Australia.

As the world's nations seek solutions to the global problem of climate change, increasing importance is being given to reducing emissions from deforestation and degradation in developing countries—the so-called REDD agenda (UNFCCC 2007). The REDD agenda is, however, not focused on forests in developed countries. From a scientific perspective, as the atmospheric warming caused by a pulse of carbon dioxide is the same regardless of the source, it is necessary to consider emissions from natural ecosystems in all countries, including Australia. Emissions from land-use activities in developed countries such as Australia are covered under the so-called 'Land Use, Land Use Change and Forestry' (LULUCF) rules in the Kyoto Protocol. The role of forests and other natural ecosystems in carbon storage is, however, presently unaccounted for by Australia unless deforestation has occurred. Therefore, new policies and measures are needed that reward land stewards for avoiding emissions from, and restoring carbon stocks in, natural forest and woodland ecosystems.

A prerequisite to good climate change mitigation policy is a reliable assessment of an ecosystem's green carbon stocks. Carbon storage in ecosystems can be assessed in terms of their current carbon stock relative to their natural carbon carrying capacity (Keith et al. 2010). Current carbon stock represents the actual carbon stock at a given time and includes the effects of the direct impacts of past disturbances due to human activities such as logging and clearing for pastoralism. Carbon carrying capacity is the amount of carbon terrestrial ecosystems can store when averaged over an appropriate time scale and area, inclusive of the landscape-level impacts of natural disturbance regimes such as fire.

The current carbon stock of terrestrial ecosystems (plants and soil) globally is at least 2000 Gt C, with about 75 per cent in forest ecosystems and 25 per cent in other ecosystem types (Houghton 2007). The international definition of forests, however, includes structural formations that in Australia are defined as woodlands and shrublands (AUSLIG 1990), such as those that characterise the GWW. We need to avoid emissions from all sources if we are to succeed in stabilising atmospheric levels of carbon dioxide at a level that prevents dangerous climate change (IPCC 2007). Protecting and restoring the green carbon stocks in ecosystems such as the GWW can contribute to solving the global climate change problem.

Human activity can lead to either an increase or a decrease in the frequency of fires and a change in fire regimes (that is, the pattern of fire events) (Gill 1975). This change in fire regimes can be considered an indirect effect of humans on an ecosystem's carbon carrying capacity. On a landscape-wide basis, forest ecosystems have been shown to be relatively resilient to fire due to a combination of factors, including: i) the existence of very large old trees that are fire resistant; ii) the presence of tree species with fire-adapted plant life history traits; iii) the occurrence of fire refugia due to landform complexity; and iv) negative feedbacks on flammability (Mackey et al. 2002). The impact of fire regimes on the carbon carrying capacity of woodland ecosystems is, however, poorly studied. As noted above, the available evidence suggests that the woodlands of the GWW could be less resilient to changes in the fire regime than are the wetter eucalypt forests of south-western and south-eastern Australia (Hopkins and Robinson 1981).

In summary, the current carbon stocks of woodland ecosystems can increase or decrease due to: 1) variation in the natural conditions that control plant growth and decay; 2) direct human impacts such as timber cutting and land clearing; and 3) indirect impacts of humans on fire regimes. By definition, carbon carrying capacity represents the carbon stock of an ecosystem inclusive of natural but not anthropogenic disturbance. Therefore, the carbon carrying capacity provides a baseline with which to compare the current carbon stock in order to assess the impacts of human activities. Such analyses make it possible to identify management options for protecting and restoring ecosystem carbon stocks at the landscape scale.

1.2 AIMS

The aims of this report are to

- quantify the current carbon stock of the GWW

- assess how this differs from the carbon carrying capacity

- identify the impacts that have reduced the carbon carrying capacity due to the direct and indirect impacts of human activities

- recommend possible management options for protecting and restoring the carbon carrying capacity of the GWW.

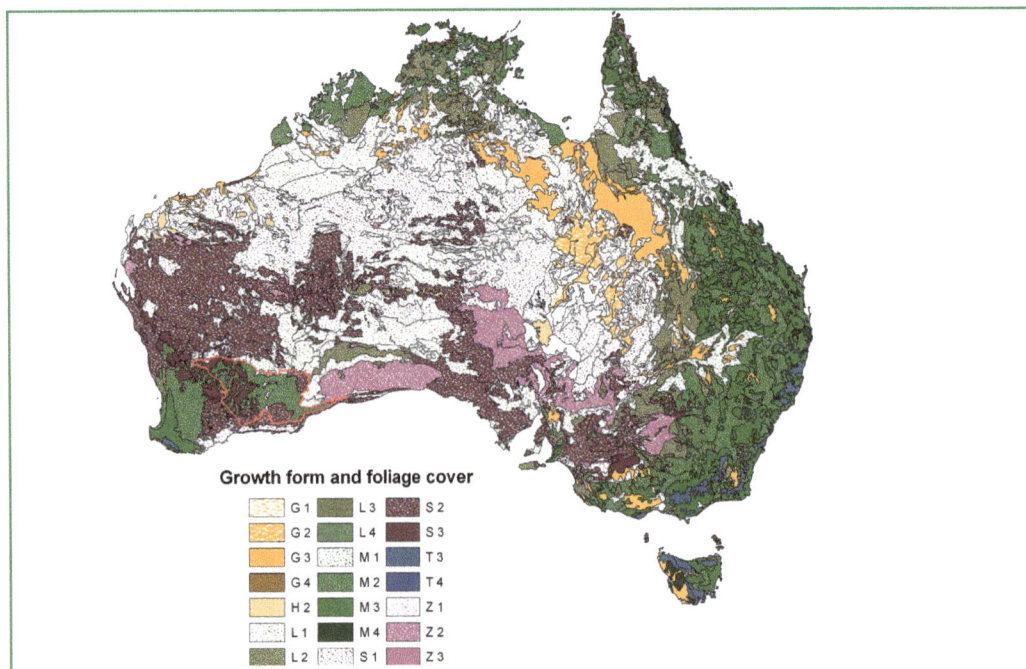

Figure1.1a

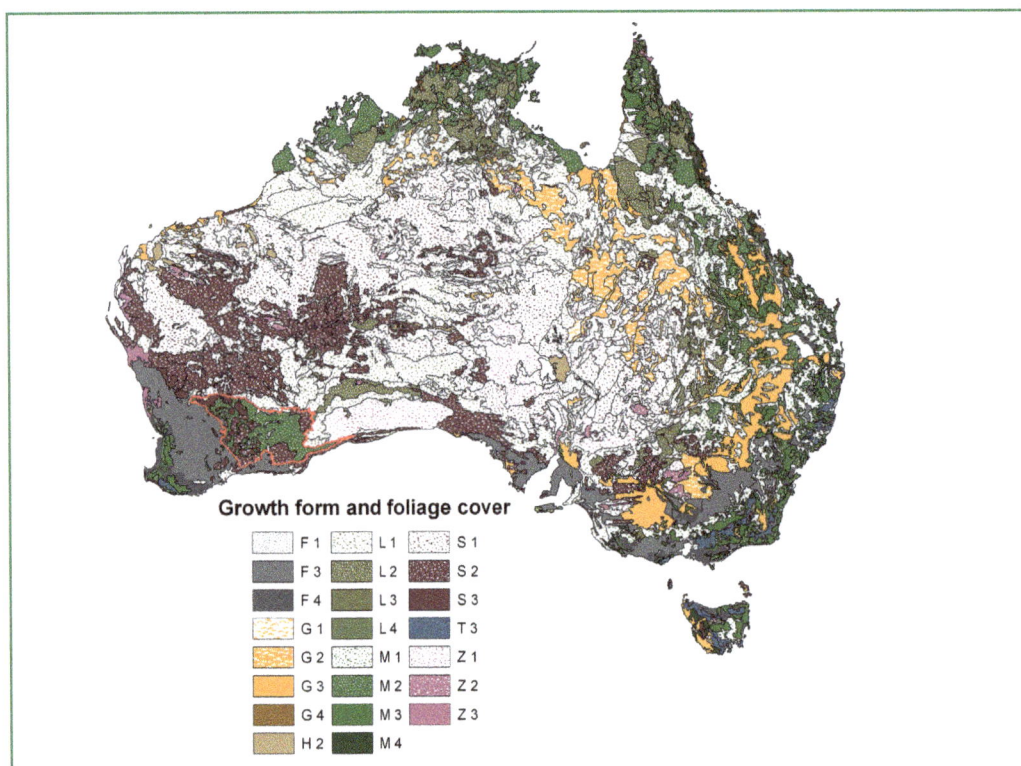

Figure 1.1b

Figure 1.1 Natural vegetation (a) and present vegetation (b) as mapped by Carnahan (AUSLIG 1990). The GWW study area is outlined in red. Vegetation coding: growth form of the tallest stratum is indicated by letters. T = tall trees (>30 m); M = medium trees (10–30 m); L = low trees (<10 m); S = tall shrubs (>2 m); Z = low shrubs (<2 m); H = hummock grasses; G = tussocky or tufted grasses and graminoids; F = other herbaceous plants. Foliage cover of the tallest stratum is classified into numeric classes: 1 = <10%; 2 = 10–30%; 3 = 30–70%; 4 = >70%.

These figures incorporate natural vegetation and present vegetation data that are Copyright Commonwealth of Australia 2003.

2. ENVIRONMENT AND HISTORY OF THE GWW

This chapter provides an overview of the topography, climate, geology, vegetation and recent land-use history—factors that play a pivotal role in determining the current carbon stock and carbon carrying capacity of the GWW.

2.1 TOPOGRAPHY

The land surface elevation of the GWW rises from ~140 m near the southern and eastern boundaries to ~500 m above sea level to the west of Kalgoorlie (Figure 2.1). The region is characterised by broad flat ridges and broad flat valleys (Figure 2.2). There are no rivers to carry water to the sea. The watercourses are shallow and drain internally into chains of salt lakes from which the water evaporates.

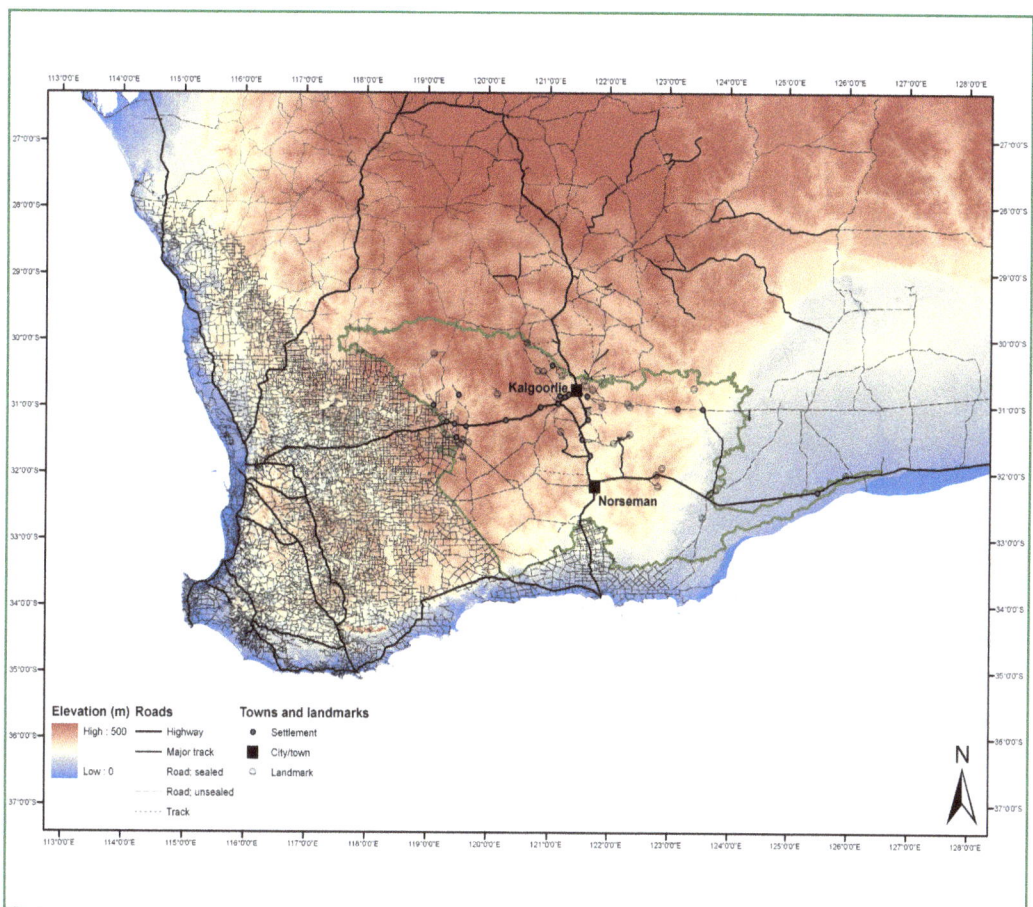

Figure 2.1 Map showing elevation (above mean sea level) of the land surface in south-western Western Australia. The GWW is outlined in green. Roads, towns and landmarks are also shown.

Spatial data sources: see Table A1—Elevation, Roads, Towns. This figure incorporates Digital Elevation Model Version 3 and Population Centre data, which are Copyright Commonwealth of Australia (2008 and 1998 respectively) and road location data that are Copyright State of Western Australia 2007.

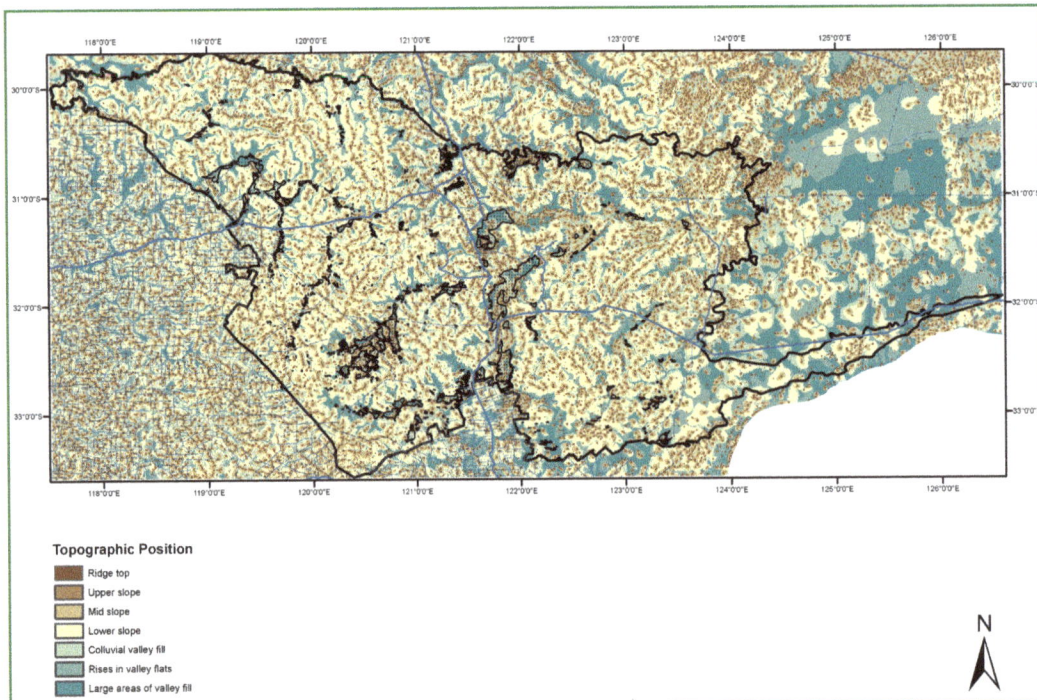

Figure 2.2 Map showing the topography of the GWW. Water courses drain into salt lakes, which are indicated by stippled patterning. Roads are shown as blue lines.

Spatial data sources: see Table A1—Topo, Roads. This figure incorporates road location data that are Copyright State of Western Australia 2007.

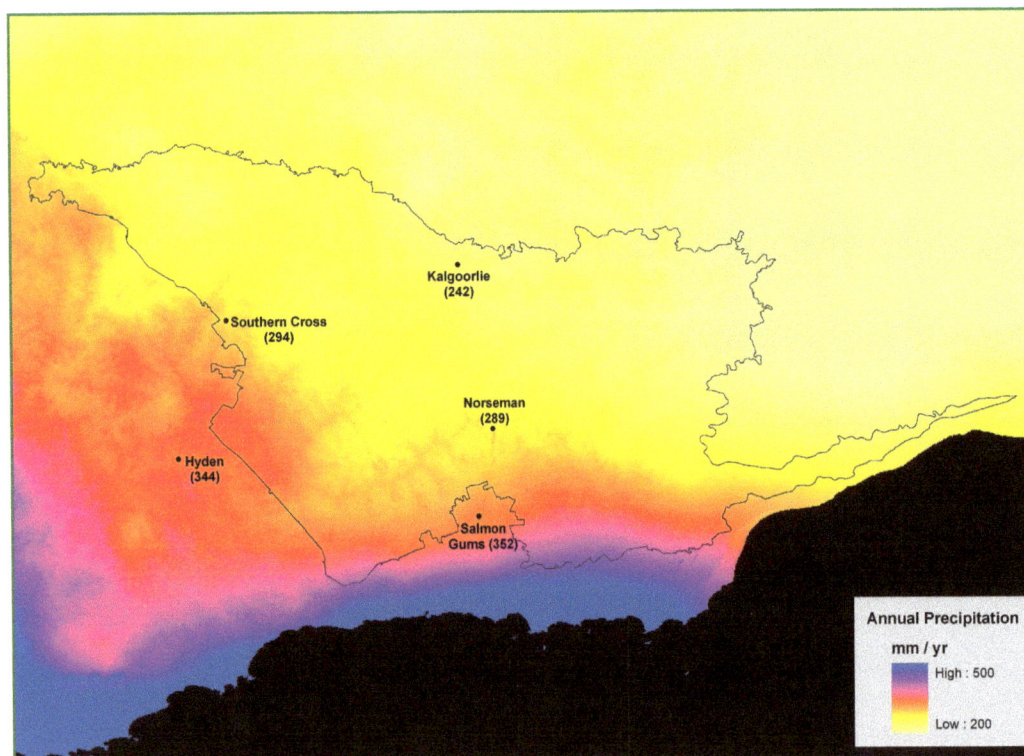

Figure 2.3 Mean annual precipitation over the GWW. Mean annual rainfall (mm) based on the historical record is given in parentheses.

Spatial data source: see Table A1—P. Mean annual rainfall source: Bureau of Meteorology (<http://www.bom.gov.au/>). This figure incorporates climate data that are Copyright Commonwealth of Australia 2009.

2.2 CLIMATE RANGE AND VARIABILITY

2.2.1 Precipitation

The rainfall across the whole of the GWW is low—ranging from approximately 350 mm yr^{-1} in the south to fewer than 250 mm in the north of the region (Figure 2.3). Rainfall can occur during any month of the year but there is more pronounced winter rainfall in the south-west, with decreasing seasonality (and total precipitation) to the north-east.

2.2.2 Temperature

The GWW experiences long, hot summers and short, frosty winters. The spatial variability in mean maximum (January) and minimum (July) temperatures is shown in Figures 2.4 and 2.5. There is a gradient in maximum temperature with distance from the coast. During the summer, sea breezes have been observed in Kalgoorlie (360 km inland) at about 9 pm (Linacre and Hobbs 1977). Closer to the coast, the sea breeze has a greater cooling effect. Southern Cross experiences an average of 113 days above 30°C each year and 11 days above 40°C. In contrast, Salmon Gums experiences 63 days above 30°C and four days above 40°C. The minimum (July) temperatures in Southern Cross and Salmon Gums are, however, very similar. Both these centres experience about 38 frosty nights a year, while Kalgoorlie and Norseman experience, on average, 22.

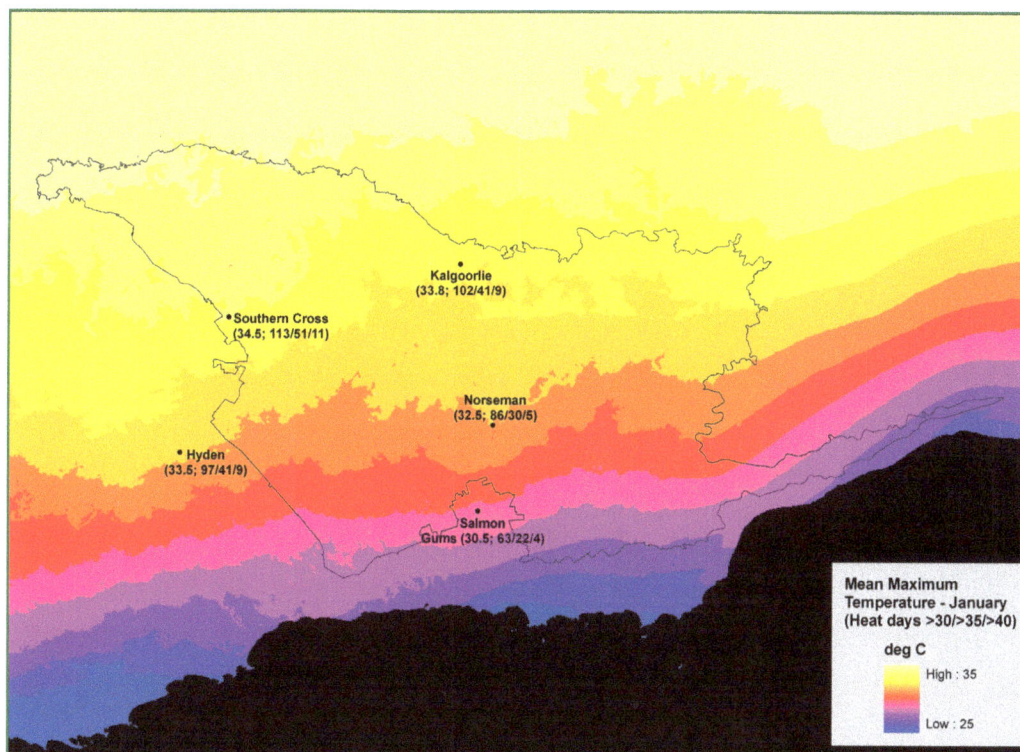

Figure 2.4 Mean maximum daily temperature (T$_{max}$, January) over the GWW. Historical average values of T$_{max}$, along with the average number of days each year when T$_{max}$ exceeds 30°C, 35°C and 40°C, are given in parentheses.

Spatial data source: see Table A1—T$_{max}$. Historical average value source: Bureau of Meteorology (<http://www.bom.gov.au/>). This figure incorporates climate data that are Copyright Commonwealth of Australia 2009.

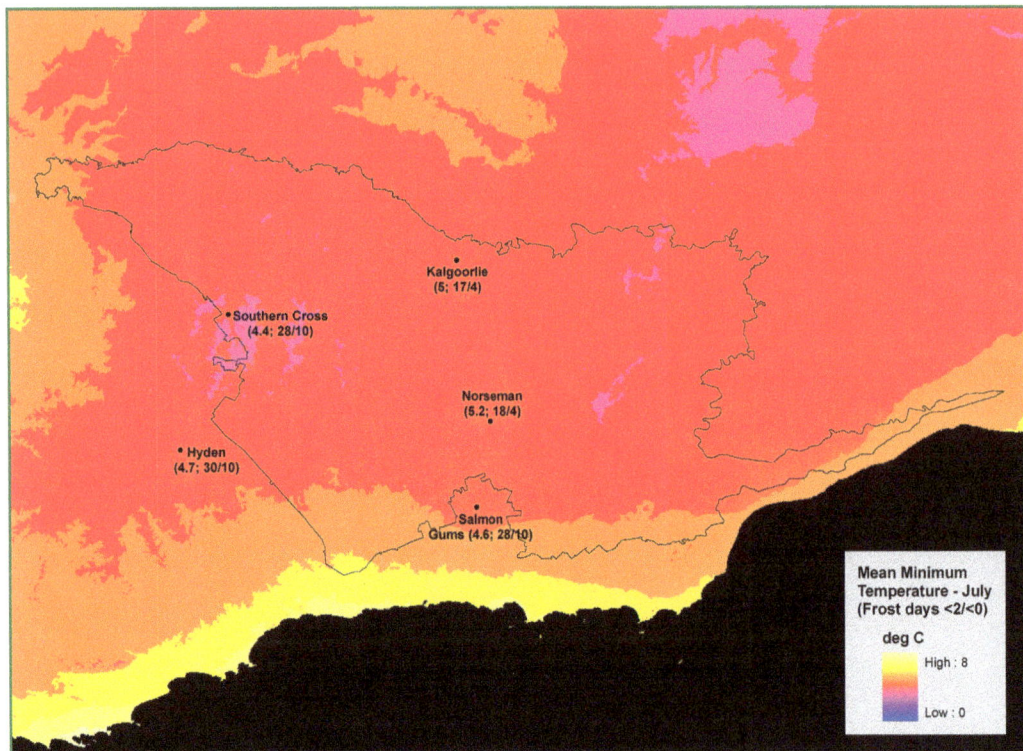

Figure 2.5 Mean minimum daily temperature (T_{min}, July) over the GWW. Historical average values of T_{min}, along with the average number of frosty nights each year when T_{min} is less than 2°C and 0°C, are given in parentheses.

Spatial data source: see Table A1—T_{min}. Historical average value source: Bureau of Meteorology (<http://www.bom.gov.au/>). This figure incorporates climate data that are Copyright Commonwealth of Australia 2009.

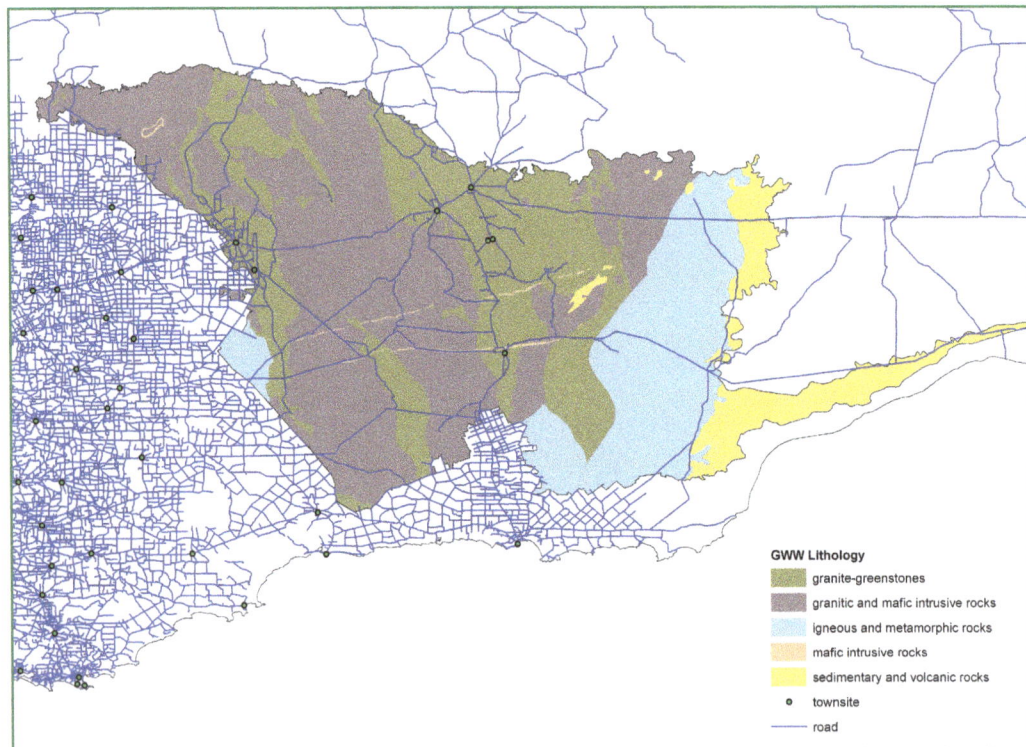

Figure 2.6 Location of major lithological units within the GWW.

Spatial data source: see Table A1—Lithology, Roads, Towns. This figure incorporates population centre data that are Copyright Commonwealth of Australia 1998, and geology and road location data that are Copyright State of Western Australia 2007.

2.3 GEOLOGY AND SOILS

The GWW lies on four major lithological units (Figure 2.6): 1) granite-greenstones; 2) granite and mafic intrusive rocks; 3) igneous and metamorphic rocks; and 4) sedimentary and volcanic rocks. Two thin bands of mafic intrusive rocks traverse the central GWW. The granite-greenstone lithological unit contains the majority of Australia's high-quality gold deposits (Henson and Blewett 2006) in addition to other heavy-metal deposits including nickel. The ancient, deeply weathered landscape has given rise to a range of soil types formed from alluvial (water-transported) and residual sediments, along with areas of exposed bare rock.

The 1:250 000 geological series of sheet maps published by the Geological Survey of Western Australia (2007) provide more detailed mapping of surface geology. (The index to these sheet maps is shown in Figure 2.7.) The surface geology depicted by these sheet maps was generally derived using a combination of field survey and aerial photo patterns, with much dependence on the latter in those areas that were inaccessible by four-wheel-drive vehicle (see Table A2).

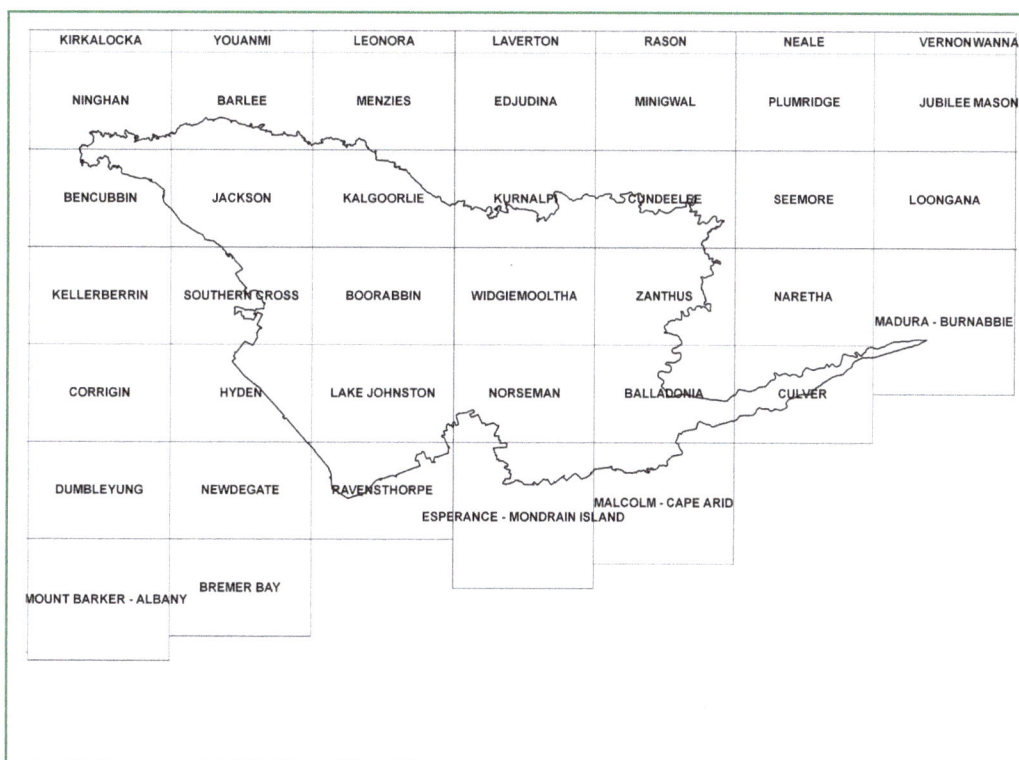

Figure 2.7 Coverage of GWW by 1:250 000 geological series map sheets.

Data sources: see Table A1—Map Sheets and Table A2 for information extracted from the Explanatory Notes that accompany the map series. This figure incorporates map sheet Index250 data that are Copyright State of Western Australia 2007.

2.4 VEGETATION

Surveys of the vegetation of the GWW region by Beard (1968, 1972, 1976) and the Western Australia Museum (Newby et al. 1984) identified six plant formations: 1) scrub heath; 2) broombush thicket; 3) rock pavement vegetation; 4) mallee; 5) sclerophyll woodland; and 6) halophytes (see Figure 2.8). The relationships between these plant formations, the dominant species and soil types are summarised in Tables A3 and A4. Soil type is an important determinant of species composition as it affects both the availability of nutrients and water storage (Brady and Weil 2002). Leached sands support low scrub heath vegetation (also known as Kwongan). Prominent species in these scrub heath shrublands belong mostly to the Proteaceae and Myrtaceae plant families, while eucalypts are absent. Broombush thickets (tall shrublands) are found on shallow, sandy soils and consist mostly of *Casuarina, Acacia and Melaleuca* species. Rock pavement vegetation includes a range of plant structures from lichens and mosses through to shrubs that occupy crevices where soil has accumulated. Mallee (multi-stemmed eucalypt) shrublands occur on both leached granite soils and deeper residual and alluvial soils. Sclerophyll (eucalypt) woodlands occur on the deeper soils. The understorey of sclerophyll woodlands is often lacking or is made up of a sparse or open shrub layer or layers or occasionally spinifex. The composition of the understorey shrubs reflects the soil type. Finally, the halophyte plant formation (consisting of succulent or semi-succulent low shrubs) is found on the highly saline depressions around salt lakes.

Distributions of plant formations and the species that constitute them cannot readily be mapped over large areas without homogenisation of finer-scale spatial heterogeneity. Consequently, various classification schemes have been developed, including the schemes used by Carnahan (AUSLIG 1990) and the National Vegetation Information System (NVIS) (DEWHA 2005) for mapping the continent's vegetation cover. Beard (1968) also applied a scheme to map the vegetation for the 1:250 000 printed map sheet series.

Natural (1780s) vegetation and present (1980s) vegetation structural formations, as mapped by Carnahan (AUSLIG 1990), are shown in Figure 2.9. The vegetation structural groups form the framework of Table 2.1. Carnahan's continental vegetation map (AUSLIG 1990) identified five vegetation structural groups in the GWW in both the present (1980s) and the natural (1780s) vegetation. More recent mapping by the National Vegetation Information System (NVIS 3.1; DEWHA 2005) identifies 14 major vegetation groups (MVGs) within the GWW (Figure 2.10). Based on that mapping, the areal extent of the MVGs in the GWW can be estimated using a Geographic Information System (see Table 2.2). We present the NVIS 'extant' vegetation layer in Figure 2.10, but not the NVIS 'pre-1750' layer. These layers differ only where vegetation has been cleared for agricultural crops. Thus, it might be assumed that the vegetation in the GWW has changed little in the past two centuries. As detailed below, however, extensive areas have been impacted on by fire, mineral exploration, pastoralism and timber cutting. Of these impacts, only timber cutting is well documented.

Figure 2.8. Major vegetation types in the GWW identified by Beard (1968).

Figure 2.8.1a Scrub heath.

Figure 2.8.1b

Figure 2.8.2a Broombush thicket.

Figure 2.8.2b

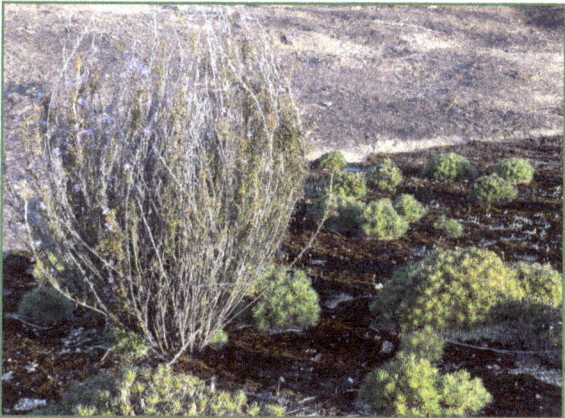

Figure 2.8.3a Rock pavement vegetation.

Figure 2.8.3b

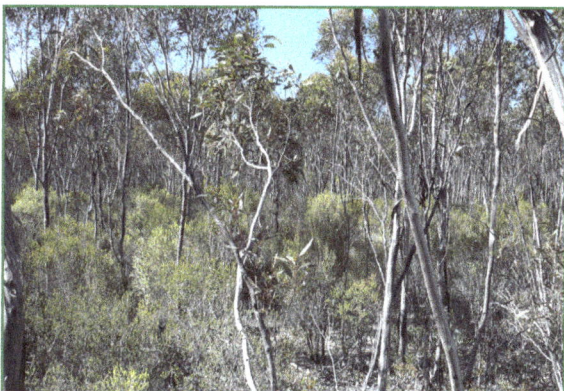

Figure 2.8.4c Mallee shrublands and woodlands.

Figure 2.8.4d

Figure 2.8.5a Sclerophyll woodlands.

Figure 2.8.5b

Figure 2.8.5c

Figure 2.8.5d

Figure 2.8.5e

Figure 2.8.5f

Figure 2.8.6a Halophytes.

Figure 2.8.6b

Photos: S. Berry

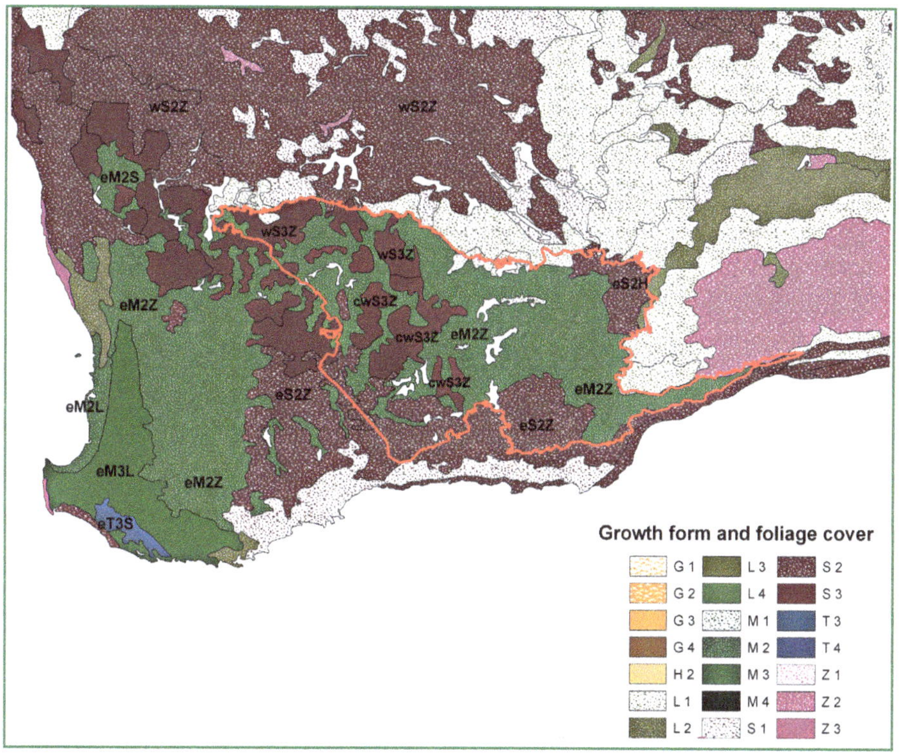

Growth form and foliage cover

G 1	L 3	S 2
G 2	L 4	S 3
G 3	M 1	T 3
G 4	M 2	T 4
H 2	M 3	Z 1
L 1	M 4	Z 2
L 2	S 1	Z 3

Figure 2.9a

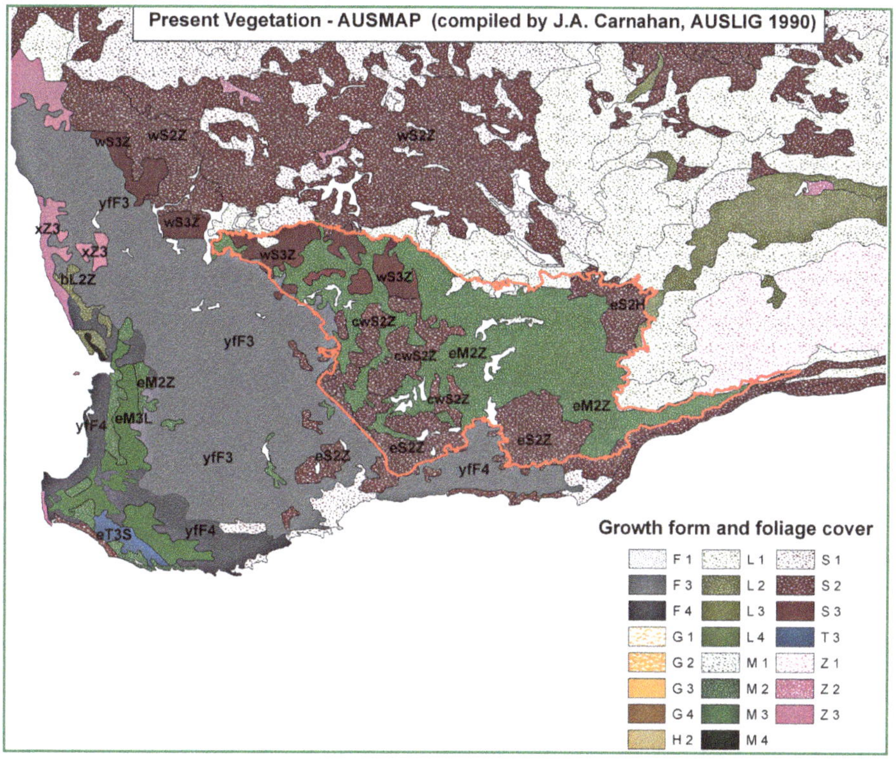

Present Vegetation - AUSMAP (compiled by J.A. Carnahan, AUSLIG 1990)

Growth form and foliage cover

F 1	L 1	S 1
F 3	L 2	S 2
F 4	L 3	S 3
G 1	L 4	T 3
G 2	M 1	Z 1
G 3	M 2	Z 2
G 4	M 3	Z 3
H 2	M 4	

Figure 2.9b

Figure 2.9 Natural vegetation (a) and present vegetation (b) as mapped by Carnahan (AUSLIG 1990). The GWW study area is outlined in red. For growth form and foliage cover codes, see Figure 1.1. The alpha-numeric codes describe the dominant structural floristic cover. The first three (or occasionally four) characters describe the tallest stratum while the remaining character refers to the growth form of the lower stratum. The lower-case letters denote the predominant floristic type: b = Banksia; c = Casuarina (including Allocasuarina); e = Eucalyptus; f = Fabaceae (including clovers and medics); w = Acacia; x = mixed or other; y = other grasses.

These figures incorporate natural and present vegetation data that are Copyright Commonwealth of Australia 2003.

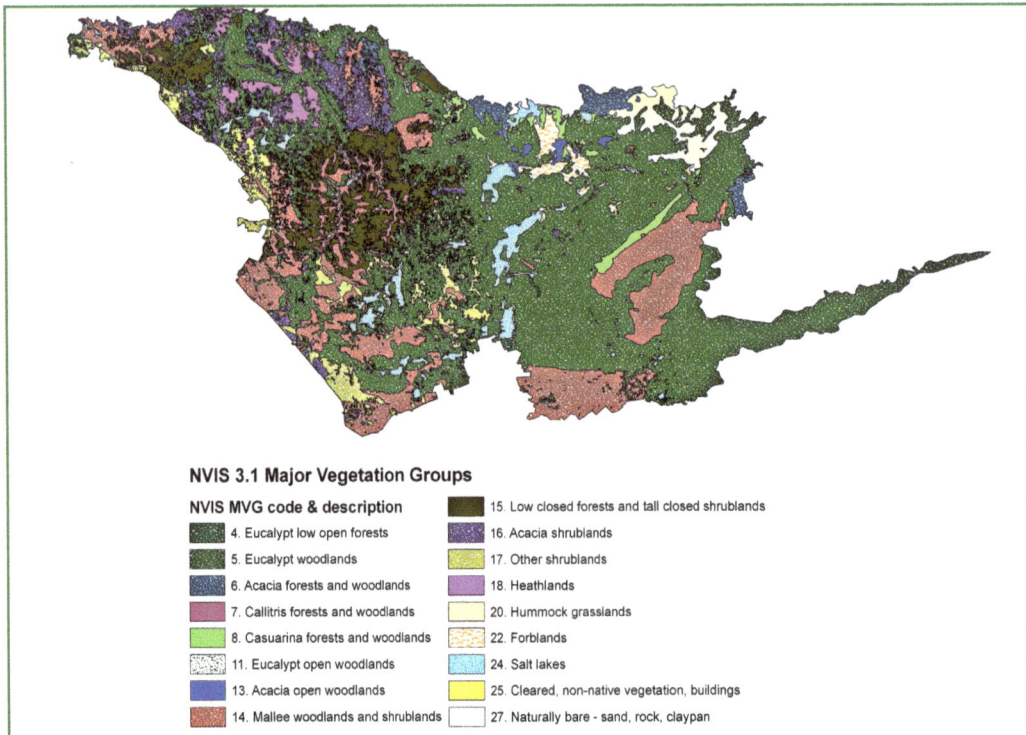

NVIS 3.1 Major Vegetation Groups

NVIS MVG code & description

- 4. Eucalypt low open forests
- 5. Eucalypt woodlands
- 6. Acacia forests and woodlands
- 7. Callitris forests and woodlands
- 8. Casuarina forests and woodlands
- 11. Eucalypt open woodlands
- 13. Acacia open woodlands
- 14. Mallee woodlands and shrublands
- 15. Low closed forests and tall closed shrublands
- 16. Acacia shrublands
- 17. Other shrublands
- 18. Heathlands
- 20. Hummock grasslands
- 22. Forblands
- 24. Salt lakes
- 25. Cleared, non-native vegetation, buildings
- 27. Naturally bare - sand, rock, claypan

Figure 2.10 Extant vegetation cover of the GWW as mapped by NVIS 3.0 (see Table A1).

This figure incorporates NVIS Version 3.0 data that are Copyright Commonwealth of Australia 2005.

We show the relationships between the classification schemes of Carnahan, the NVIS and Beard in Table 2.1. Although the schemes all classify the vegetation according to canopy height, canopy cover and floristic association of dominant taxa, the derived classes cannot be readily equated across schemes.

Table 2.1 Relationship between the classification schemes used by Carnahan (AUSLIG 1990) shown in bold, NVIS MVGs (DEWHA 2005) and Beard (1968) shown in italics. Classes with green shading do not occur in the GWW.

Canopy height (m) and height class		Projective foliage cover of tallest stratum (PFC; %)			
		100–70	70–30	30–10	<10
10–30	M	**M4—closed forest**	**M3—open forest** MVG 3	**M2—woodland** MVG 5, 6, 7, 8, 14 *Sclerophyll woodland*	**M1—open woodland** MVG 11, 14
5–10	L	**L4—low closed forest** MVG 15	**L3—low open forest** MVG 4, 6, 7, 8, 14, 15?	**L2—low woodland** MVG 5, 6, 7, 8, 14	**L1—low open woodland** MVG 11, 13, 14
2–5	S	**S4—closed shrub** MVG 15 *Broombush thicket Mallee*	**S3—open scrub** MVG 14, 15?, 16, 17, 18 *Broombush thicket Mallee*	**S2—tall shrubland** MVG 14	**S1—tall open shrubland** MVG 14
<2	Z	**Z4—closed heath** MVG 15, 18 *Scrub heath*	**Z3—open heath** MVG 16, 17, 18, 22 *Scrub heath*	**Z2—low shrubland** MVG 18, 22 *Halophytes*	**Z1—low open shrubland** MVG 18, 22 *Halophytes*
<1	H			**H2—hummock grassland** MVG 20	
<1	F			**F2—open herbfield**	**F1—sparse open herbfield** *Rock pavement vegetation*

Table 2.2 Areal extent of cover of NVIS Major Vegetation Groups (version 3.0) in the GWW.

Major vegetation group	Area (× 1000 ha)
4. Eucalypt low open woodlands	4.4
5. Eucalypt woodlands	8 377.1
6. Acacia forests and woodlands	299.7
7. Callitris forests and woodlands	4.6
8. Casuarina forests and woodlands	227.9
11. Eucalypt open woodlands	15.5
13. Acacia open woodlands	64.2
14. Mallee woodlands and shrublands	2 748.3
15. Low closed forests and tall closed shrublands	1 048.7
16. Acacia shrublands	748.6
17. Other shrublands	527.8
18. Heathlands	304.4
20. Hummock grasslands	332.4
22. Forblands	227.3
24. Salt lakes	616.4
25. Cleared, non-native vegetation, buildings	222.1
27. Naturally bare—sand, rock, claypan, mudflat	202.4
Total native vegetation cover	14 931.0
Total area of GWW	15 972.0

2.5 TIMBER CUTTING IN THE GWW

The timber resource of the GWW was pivotal to the development of the Eastern Goldfields. Between 1897 and 1969, the woodlands were cut to provide 22 million tonnes of timber and fuel wood for the mines of the Golden Mile (Kalgoorlie region), mostly to fuel the boilers to power mining machinery (Figures 2.11–2.13) (Western Australia Forests Department 1969). Additionally, the woodlands were exploited to provide timber and fuel wood to the Norseman goldfields, 2.2 million railway sleepers, 3.4 million fence posts, 84 000 bean sticks and 110 000 posts, poles and bridge timbers (Western Australia Forests Department 1934–70). Salmon gum (*Eucalyptus salmonifloia*) and gimlet (*Eucalyptus salubris*) were the preferred species for timber cutting in the GWW. 'Woodlines'—railway lines—were constructed to transport the timber to the mines in Kalgoorlie. Wood was also cut to fuel—continuously for 55 years—the six pumping stations of the Goldfields Water Supply Scheme between Cunderdin and Coolgardie. Approximately one-quarter of the GWW (4.4 million ha; see Table 2.3) was cut over (estimated from the map shown in Figure 2.12), producing, on average, 10 tonnes of firewood per cut-over hectare (Western Australia Forests Department 1969). These figures indicate that the total eucalypt timber harvest could have been about 40 million tonnes. Sandalwood was also cut to provide raw materials for the production of incense, soap and perfumes (Western Australia Forests Department 1934–70).

Figure 2.11a

Figure 2.11 b

Figure 2.11c

Figures 2.11 Timber cutting in the GWW was carried out principally by the Kalgoorlie Boulder Firewood Company. Train loaded with logs (a); Horse-drawn carts stacked with wood, 1910. (b); and logs stacked by the woodline (c).

Photos reproduced by permission of The Battye Library, State Library of Western Australia (image reference numbers: 005703D, 005705D, 005708D).

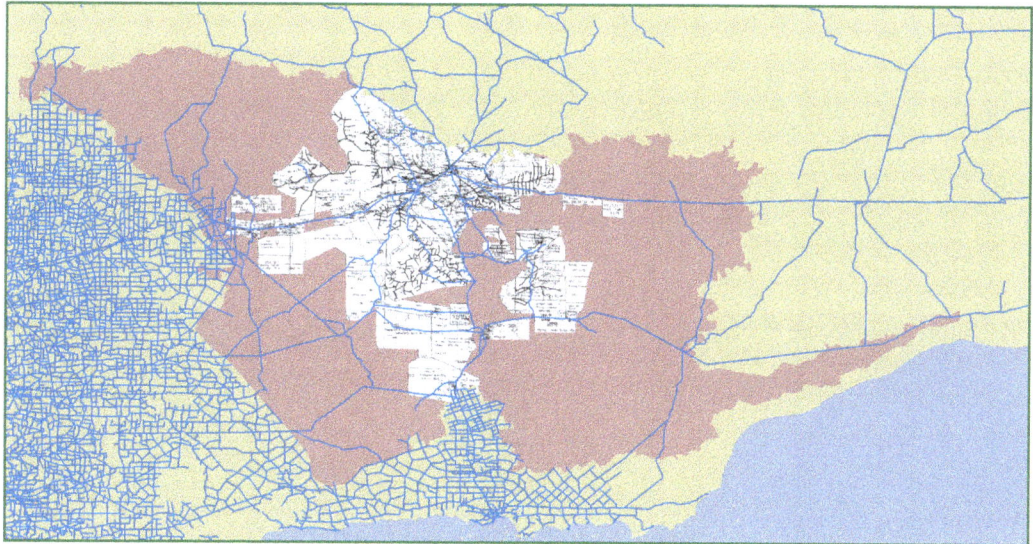

Figure 2.12 Map showing the extent of timber cutting in the GWW. The area within the GWW that was not subjected to timber cutting is coloured pink. The timber-cut area is shown as a transparent overlay over the 1:50 000 scale map 'Timber tramlines and cutting areas in the Goldfields Region 1900–1975' (FD No. 1610, 1978), which identifies the location of woodlines and coupe boundaries. Roads (see Table A1) are shown as blue lines.

This figure incorporates road location data that are Copyright State of Western Australia 2007.

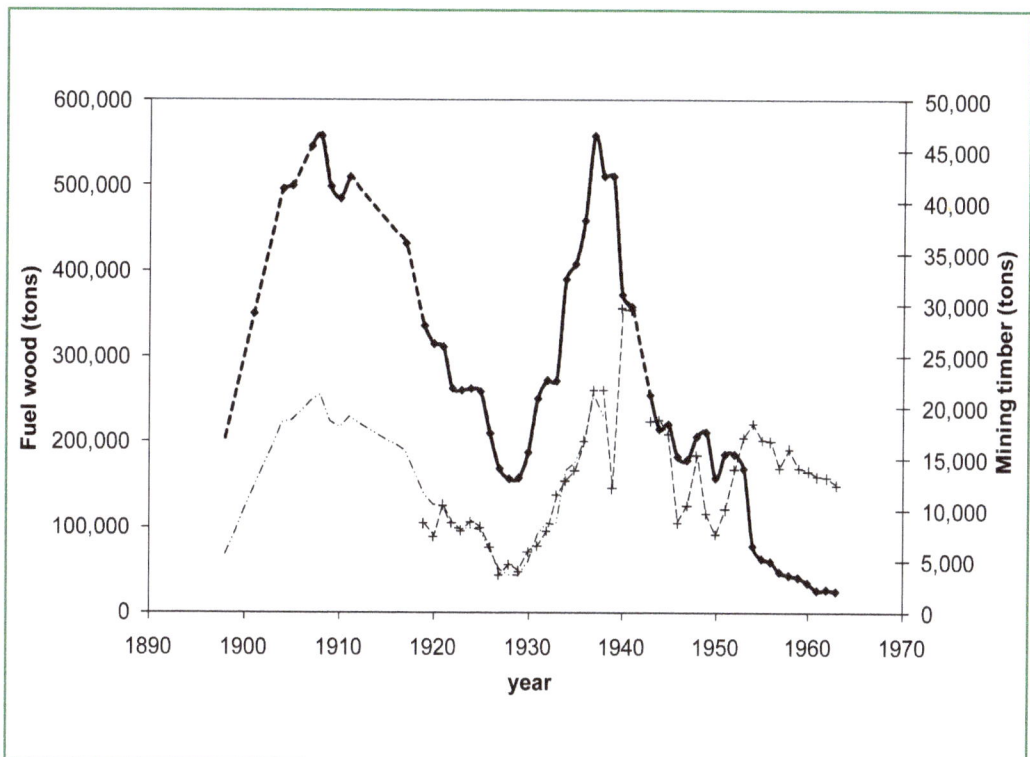

Figure 2.13 Timber harvested (tons) for fuel wood and mining from public land in the Eastern Goldfields in the period 1898–1963. Data from Forests Department reports are shown in solid lines with black diamonds (fuel wood) and dashed lines with '+' (mining timber). To estimate the total harvest of fuel wood, a linear interpolation between the reported values was used (shown by the dashed line). Before 1938, there is a linear relationship between fuel wood and mining timber (mining = 0.0438; fuel wood = 3245.8; $R^2 = 0.92$) and this is used to estimate the mining timber consumption pre-1939 (also shown as dashed line). Note: 1 ton is equivalent to 1.016 tonne.

Data from Forests Department reports compiled by Bianchi et al. (2008).

Table 2.3 Area within current (1993) land tenure types that was utilised for fuel wood and mining timber between 1900 and 1975.

Land tenure type	Estimated area affected by timber cutting between 1900 and 1975 (ha × 1000)	Total area (ha × 1000)
1. Aboriginal reserve	0	108
2. Aboriginal leasehold	22	252
3. Forestry reserve	38	120
4. Other freehold	10	249
5. Other leasehold	1 598	3 513
6. Mixed	9	9
7. Nature conservation reserve	155	1 651
8. Other	91	160
9. Vacant crown land	2 402	9 859
10. Water reserve	42	44
Total	4 367	15 965

Spatial data sources for calculations: see Table A1—Land tenure, timbercutting.

2.6 OVERVIEW OF ANTHROPOGENIC IMPACTS IN THE GWW

With the exception of the freehold land (cleared for cropping) on the western boundary, the GWW retains a cover of native vegetation. About half of the GWW, however, has been subjected to direct anthropogenic impacts of mineral exploration and mining, timber cutting and pastoralism (Figure 2.14) that have resulted in the thinning or removal of biomass. As discussed in Chapter 3, a consequence of these impacts is that the current carbon stock will be below the natural carbon carrying capacity. In Chapter 3, we also quantify the extent of another major agent of disturbance in the GWW: recent intense wildfire.

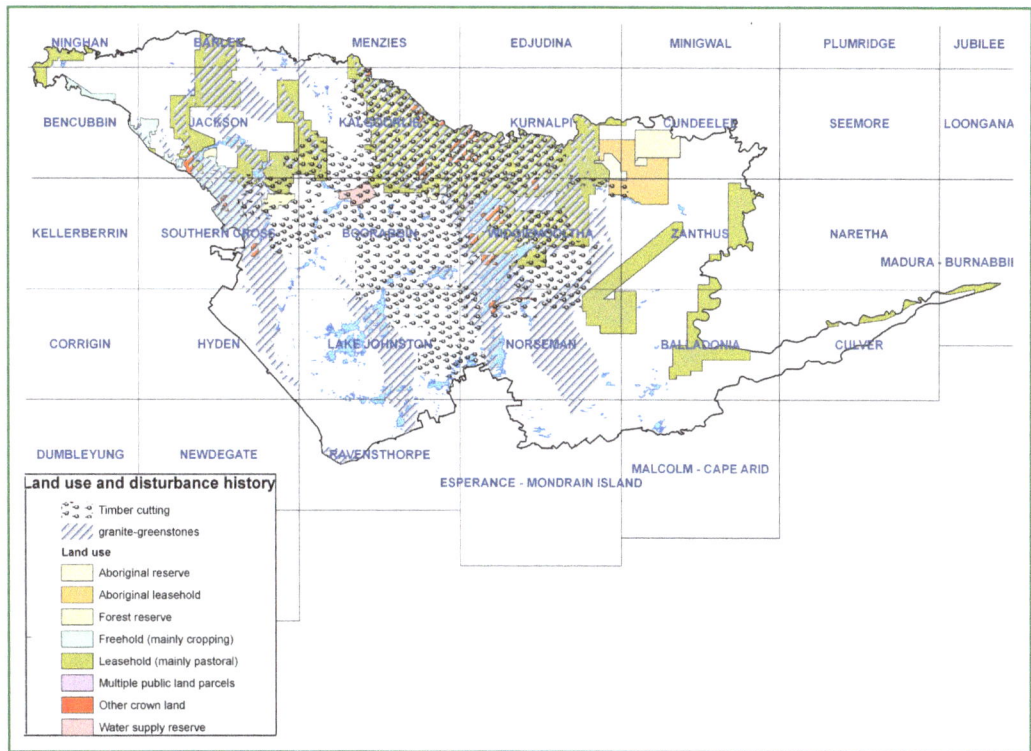

Figure 2.14 Map showing area affected by timber cutting, mining and mineral exploration (indicated by the granite-greenstone lithology) and land-use change in the GWW. The index of 1:250 000 geological series map sheets is shown as an overlay.

Reference: see Table A1—Map Sheet, Land Tenure, Lithology and Table A2. This figure incorporates Australian land tenure (1993) data that are Copyright Commonwealth of Australia 1993, and map sheet Index250 and geology data that are Copyright State of Western Australia 2007.

3. METHODS TO CALCULATE CARBON STOCKS

Two groups of data were used to develop a model of carbon stocks—namely: 1) site-specific measurements obtained through field survey; and 2) spatial data layers derived from GIS, remote sensing and environmental modelling. Site-specific data include inventory plots with measurements of the number and dimensions of plants (usually stem diameter and height) within a defined area. Allometric equations were used to estimate biomass from the measurements of plant dimensions. To extrapolate carbon stocks calculated from site-specific measurements to other locations requires spatial data layers that describe the spatial variation in environmental factors that influence biomass. These data layers (see Table A1) describe the vegetation, climate, geology and land-use history factors of mining exploration, timber cutting and wildfire occurrence. Correlations in variation of the spatial environmental factors and biomass at the field sites form the algorithms used in the extrapolation model of carbon stocks.

3.1 SITE-SPECIFIC DATA

3.1.1 Field data

ANU researchers obtained site inventory data of trees and shrubs during three field surveys from 2005 to 2007. Data were obtained for 21 sites of 250 m × 250 m in area (6.25 ha). The large sites were used: 1) to obtain representative measurements of tree and shrub size distributions in heterogeneous communities; and 2) to correspond to the pixel size of remote sensing imagery, and thus reduce the errors associated with up-scaling from small plots to pixels. Sites were marked out such that the vegetation within the boundaries was as homogeneous as possible. Site locations are shown in Figure 3.1.

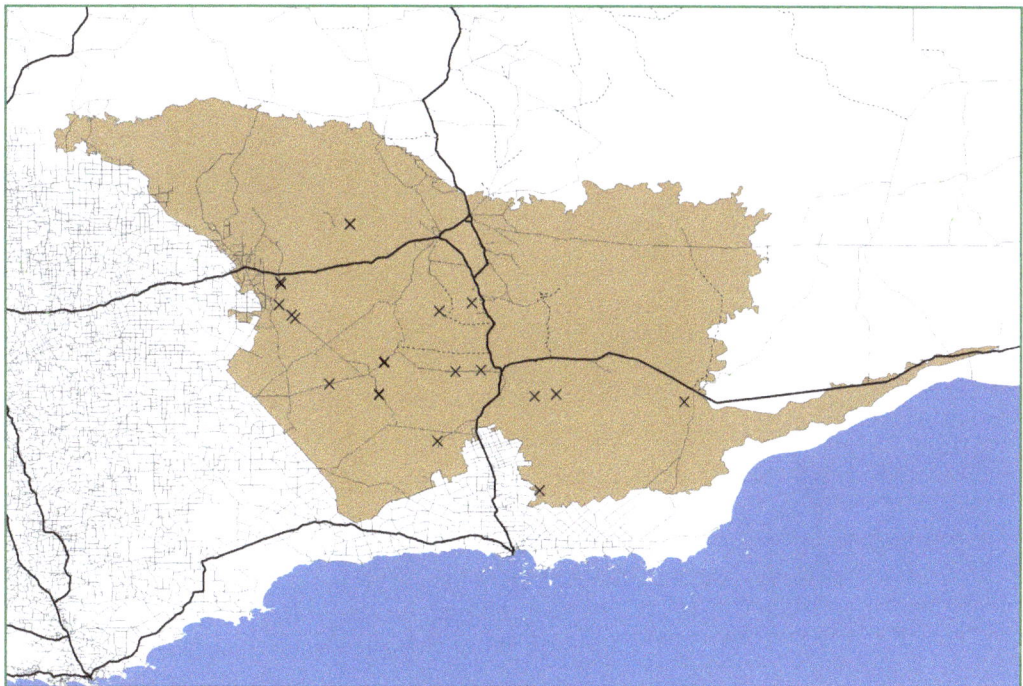

Figure 3.1 Map of the GWW showing the location of the 21 field survey sites (indicated by×). Roads (see Table A1) are shown as black lines.

This figure incorporates road location data that are copyright State of Western Australia 2007.

At each site, the major canopy layers were identified—namely: 1) monopodial or single-stemmed trees with up to three canopy layers; 2) mallee or multi-stemmed low-tree layer; 3) tall shrub layer; 4) saltbush/bluebush low shrub layer; 5) mallee shrub layer; and 6) sclerophyllous (heath) shrub layer. A Point Centred Quarter (PCQ) method was used for sampling (see Bonham 1989; Tongway and Hindley 2004). Within sites, 16 point centres were located using a grid, as shown in Figure 3.2. Measurements were made within a 25 m radius of each point centre. Within each of the four quarters surrounding a point centre, the distance was measured to the nearest stem within each canopy layer (Figure 3.3). Plant dimensions of stem diameter at 1.3 m height (corresponding to diameter at breast height, dbh) for the tree canopy layer and plant height for all canopy layers were also measured. Absolute density of plants within each canopy layer (units: stems ha^{-1}) was calculated from the number of observations and distance (see Mitchell 2007; Warde and Petranka 1981). The field data and calculations are summarised in the Appendix and Table A5.

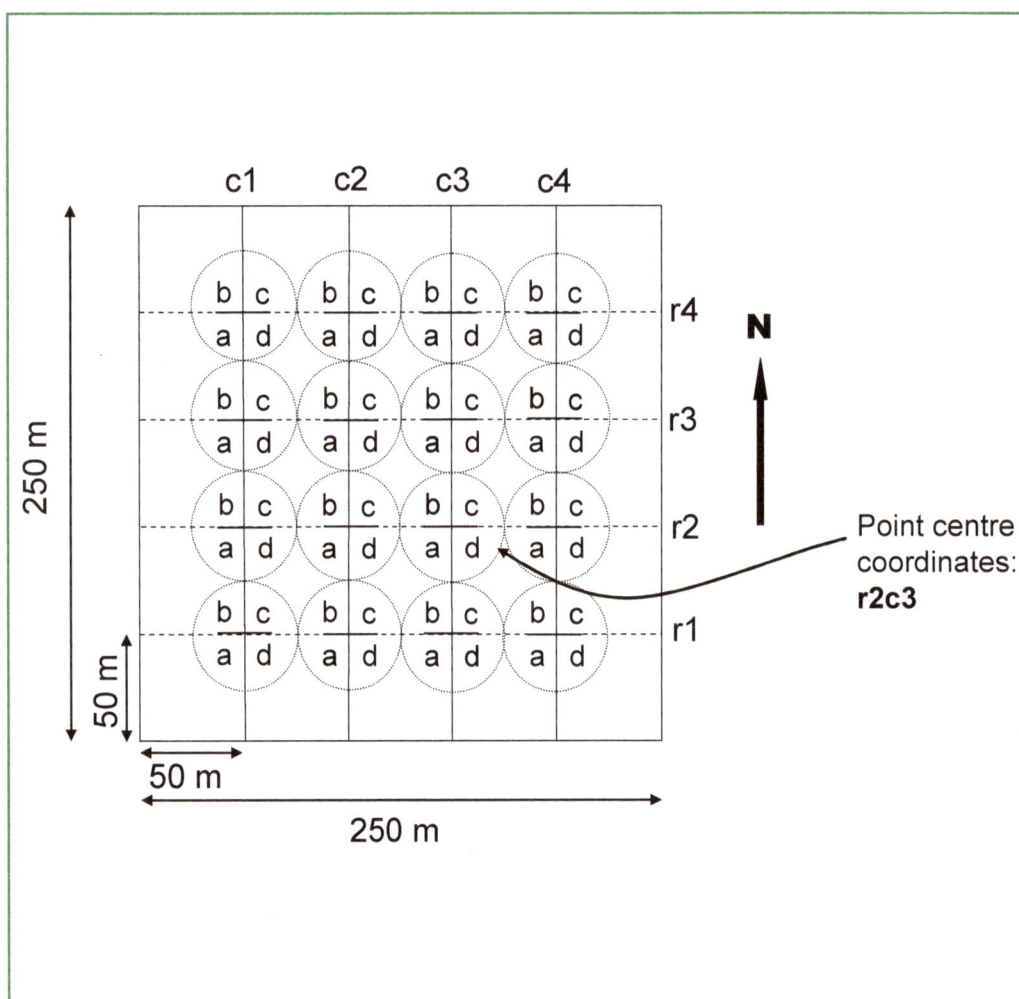

Figure 3.2 Location of the 16 point centres within a study site.

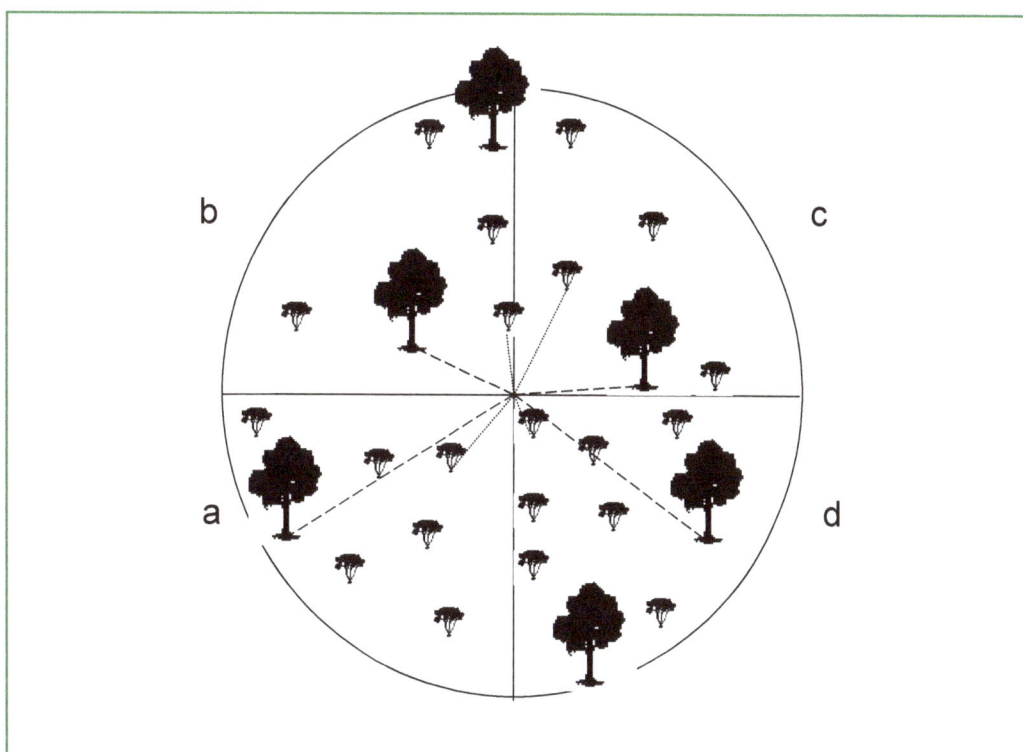

Figure 3.3 The PCQ method for measurement within each quarter (a–d) is shown as the distance from the point centre to the nearest stem of each canopy layer. Two canopy layers are shown in this example.

3.1.2 Allometric equations

Allometric equations relate tree dimensions to biomass and are derived using data from trees that were destructively sampled and the biomass components weighed. Allometric relationships describe plant morphology and vary in relation to site environmental conditions and plant genetics. Allometric equations do not exist for the species within the GWW. Hence, we compared allometric equations developed for species elsewhere in Australian woodlands to assess their utility for biomass estimation for tree and shrub canopy layers in the GWW.

A compilation of allometric equations derived for woodland species in Australia is attached in the Appendix (Table A6). Additionally, we had access to more recently derived equations for some woodland trees and mallees in south-western Western Australia (J. Jonson unpublished; see Table A7). Here, we present comparisons of predicted above-ground biomass (AGB) for eucalypt trees using generalised equations. We tested these equations using our inventory data that have a tree diameter at breast height (dbh) ranging from 2 to 118 cm. The allometric equations can be subdivided into two groups: those that require inputs of 1) stem diameter or basal area, and 2) stem diameter and stem height. An assumption of (1) is that there is a consistent relationship between stem diameter and stem height. The following are the main allometric equations for woodland trees in Australia.

Burrows et al. (2000) measured biomass of stands of *Eucalyptus crebra, E. melanophloia and E. populnea* in central Queensland rangelands. They derived the general relationship (Equation 3.1).

Equation 3.1

$AGB = 6.51B - 6.65$ (t ha^{-1})

In Equation 3.1, B is the stand basal area (m^2 ha^{-1}) (derived from diameter at 1.3 m height; dbh range 1.5–76.4 cm). This equation yields negative values of AGB for sites where the stand basal area is less than 1.0215 m^2 ha^{-1}. This problem can be circumvented by forcing the regression through zero using the simplified Equation 3.2.

Equation 3.2

$AGB = 6.10B$ (t ha^{-1})

Estimated site biomass of the tree canopy layer, based on Equations 3.1 and 3.2, is presented in Table 3.1. Using these equations, it is assumed that AGB is linearly related to stem basal area—that is, that stem height × wood basic density has a constant value (6.1 t ha^{-1}). It is, however, difficult to account for tree damage that is common in natural stands—for example, due to fire and wind—using a simplified linear relationship.

For individual *Eucalyptus melanophloia* (silver-leaved ironbark) sampled at two sites (380 km distant) in central Queensland, Burrows et al. (2001) derived Equation 3.3.

Equation 3.3

$\ln(AGB) = 2.700\ln C - 6.434$ (kg tree^{-1})

In Equation 3.3, C is the stem circumference at 30 cm above ground (dbh range 1.9–52.5 cm). The basic density[6] of this species is ~1080 kg m^{-3} (Boland et al. 1984:532)—a value similar to the basic density expected for GWW eucalypts.

For mallee eucalypts (*Eucalyptus socialis/E. dumosa* combined), at the location 33°53'S, 146°30'E (dbh range 2.5–55.2 cm), Burrows et al. (see Eamus et al. 2000 and Table A6.1) derived Equation 3.4.

Equation 3.4

$\ln(AGB) = 2.262\ln C - 4.1671$ (kg tree^{-1})

The general equation developed by Jonson (unpublished) for eucalypt woodland species (dbh range 2.3–79 cm) in the wheatbelt of south-western Western Australia (see Table A7) has the same form as Equations 3.3 and 3.4. The equation of Jonson, however, predicts AGB up to ~20 per cent smaller than Equations 3.3 and 3.4 (Figure 3.4) for the maximum stem dbh measured at our field sites.

Williams et al. (2005) have derived a general allometric equation for estimating AGB of trees (mostly eucalypts) in tropical and subtropical eucalypt woodlands across the Northern Territory, Queensland and New South Wales (Equation 3.5).

Equation 3.5

$\ln(AGB) = -2.0596 + 2.1561\ln(D) + 0.1362(\ln(H))^2$ (kg tree^{-1})

In Equation 3.5, D is the dbh at 1.3 m (dbh range 3–86 cm) and H is the tree height (m). Williams et al. found that there was good (1:1) agreement between observed and predicted values of AGB following the back-transformation of the logarithm. This implies that no correction factor needs to be applied to adjust for bias. Although the Jonson equation does not include H, there is good agreement between the predictions of

---

6   The basic density of wood is the ratio of dry mass to fresh volume.

AGB by the Jonson and Williams equations (Figure 3.5). In Figure 3.5, there is some scatter of points around the regression line because the relationship between dbh and height is not consistent in the field data set. That is to be expected because our field data include a wide range of species and age classes and many stems have suffered damage from fire, timber cutting, wind and termites. The GWW inventory data included measurements of height and dbh, so Equation 3.5 was used to estimate *AGB* of trees having dbh of 10 cm or more at our field sites.

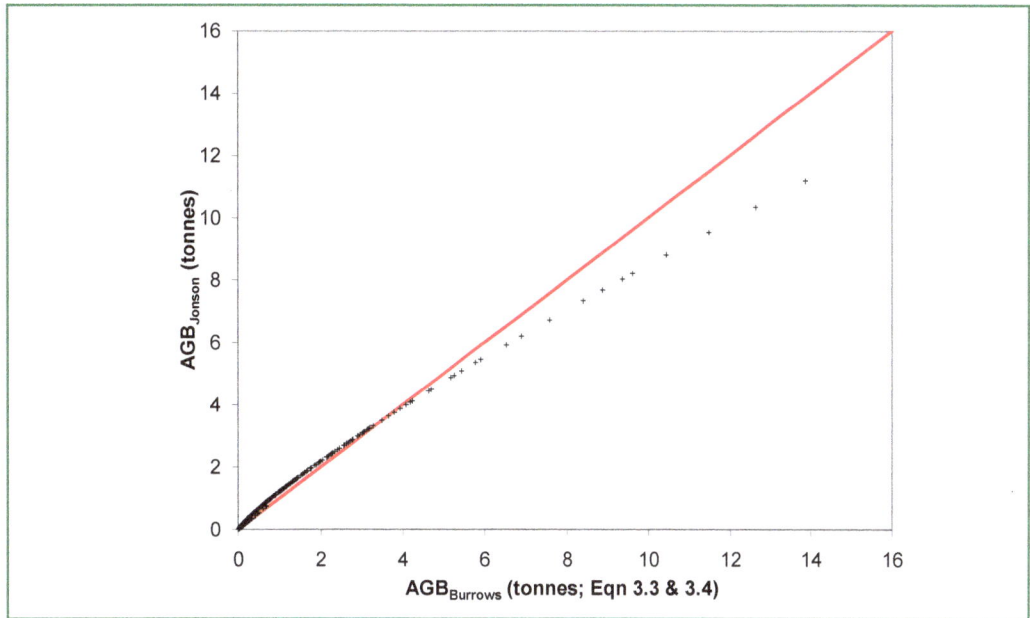

Figure 3.4 Comparison of AGB predictions from stem diameter (dbh) using the general allometric equation of Jonson and the combined equations (Equation 3.3 and 3.4) of Burrows et al. The data points are calculated AGB values (tonnes per tree) from field measurements of dbh of individual trees. The red line indicates a 1:1 relationship.

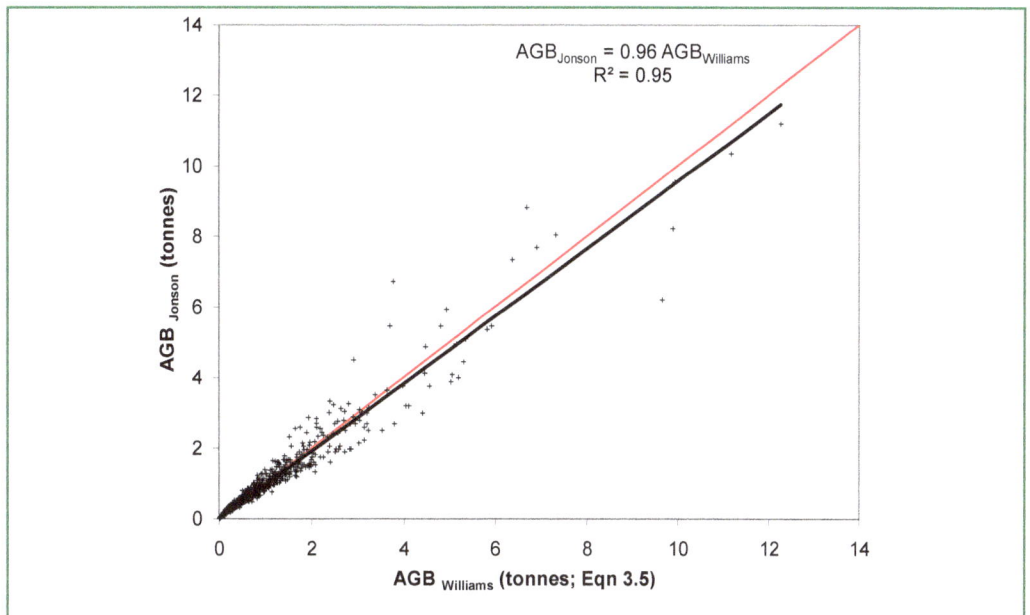

Figure 3.5 Comparison of AGB (tonnes per tree) predicted for GWW trees using Equation 3.5 (Williams et al.) and the general equation derived by Jonson. The regression line is drawn in black. The 1:1 line is drawn in red.

Table 3.1 Estimates of above-ground biomass (AGB, t ha-1) for GWW field sites. The mixed model AGB comprises the combined estimates of tree and shrub AGB from Equations 3.5 and 3.7.

Site no	Survey ID	Field vegetation description	Stand (tree) basal area (m²ha⁻¹)	AGB Burrows Equation 3.1 (t ha⁻¹)	AGB Burrows Equation 3.2 (t ha⁻¹)	Mixed model AGB (t ha⁻¹)
1	2005_01	Eucalypt woodland	4.90	25.24	29.88	46.47
2	2005_02	Open eucalypt woodland	1.48	2.98	9.02	6.14
3	2005_03	Low shrubland	no data	-	-	0.57
4	2005_04	Low shrubland	no data	-	-	2.37
5	2005_05	Eucalypt woodland	8.40	48.02	51.22	36.11
6	2005_06	Eucalypt woodland	6.04	32.70	36.87	47.51
7	2005_07	Eucalypt woodland	8.56	49.10	52.24	48.75
8	2005_08	Tall shrubland	no data	-	-	2.01
9	2005_09	Eucalypt woodland	4.34	21.58	26.45	31.80
10	2005_10	Tall shrubland	no data	-	-	5.00
11	2006_01	Low eucalyptus woodland	1.42	2.60	8.67	10.95
12	2006_02	Eucalypt woodland	7.59	42.75	46.29	57.63
13	2006_03	Eucalypt woodland	11.89	70.78	72.55	47.28
14	2006_04	Eucalypt woodland	5.74	30.71	35.01	49.10
15	2006_05	Eucalypt woodland	5.40	28.50	32.94	38.33
16	2006_06	Tall acacia shrubland	0.07	−6.18	0.44	4.35
17	2006_07	Tall acacia shrubland	0.33	−4.49	2.02	6.05
18	2007_01	Mallee shrubland	3.22	14.28	19.61	22.86
19	2007_02	Eucalypt woodland	5.30	27.83	32.30	38.19
20	2007_03	Eucalypt woodland	10.36	60.82	63.22	89.96
21	2007_04	Heath/mallee shrubland	no data	-	-	2.12

Bejan et al. (2008) have proposed that tree stems and branches have a near-conical shape, related to the mechanics of structural support and water flow through the stem and branches. In that case, there could be a simple relationship between AGB and the volume of a cone or cylinder (the volume of a cone is equal to one-third of the volume of a cylinder). For our field data set of eucalypt trees, AGB predicted by Equation 3.5 is approximately equivalent to Equation 3.6.

Equation 3.6

$AGB_{cyl} = \pi r^2 h 0.56$ (kg tree^{-1})

In Equation 3.6, r is the stem radius at 1.3 m, h is the height to the top of the canopy and the units are metres. The comparison between AGB_{cyl} and AGB predicted from Equation 3.5 is shown in Figure 3.6. We use this simple relationship to estimate the AGB of the shrub layer.

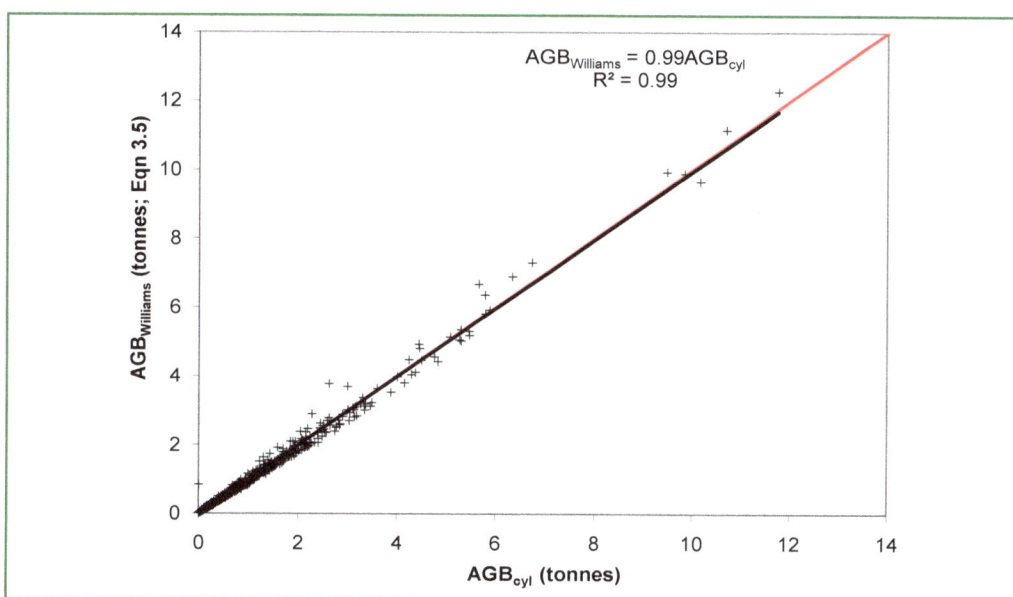

Figure 3.6 AGB (tonnes per tree) estimated from eucalypt tree and mallee measurements from our site data. The allometric equation of Williams developed for northern Australian eucalypts yields very similar estimates to those derived from the formula for a truncated cone (0.56ϖr^{2}h, in which the constant was derived from calibration with Equation 3.5).

Estimates of forest biomass commonly ignore the contribution of the shrub layer; however, about 6 million ha of the GWW has an upper canopy layer of low trees or shrubs (canopy height classes L, S and Z in Table 2.1). We measured canopy dimensions and height but not dbh for many trees with dbh <10 cm and for almost all of the shrubs recorded in our field data. Consequently, none of the allometric equations requiring a stem diameter measurement was useful for estimating shrub biomass. Allometric equations, which utilise canopy dimensions to predict arid and semi-arid shrub biomass, are considered inadequate for sites distant from the locations from which the equations were derived (Dean and Eldridge 2008). (The equations in that study yielded biomass estimates that differed by 85 per cent.) An equation that estimates AGB from height at the top of the canopy was required. We could find no suitable published equations (the equations of Harrington 1979 relate biomass to height but the species and form of the shrubs are different to those at the GWW). Instead, we derived an equation relating AGB to height for 114 plants (mostly eucalypts) that have dbh <10 cm for which we recorded a dbh measurement. First, we calculated AGB using: 1) the general allometric equation of Jonson; 2) Equation 3.4; 3) Equation 3.5; and 4) Equation 3.6. Of these equations, the best predictor of AGB from height (m) was Equation 3.6 (see Figure 3.7 and Equation 3.7).

Equation 3.7

$$AGB = 0.0002h^{2.4071} \text{ (kg plant}^{-1})$$

This equation was subsequently used to estimate the AGB of shrubs at each field site. The summed AGB for all canopy layers is presented in Table 3.1.

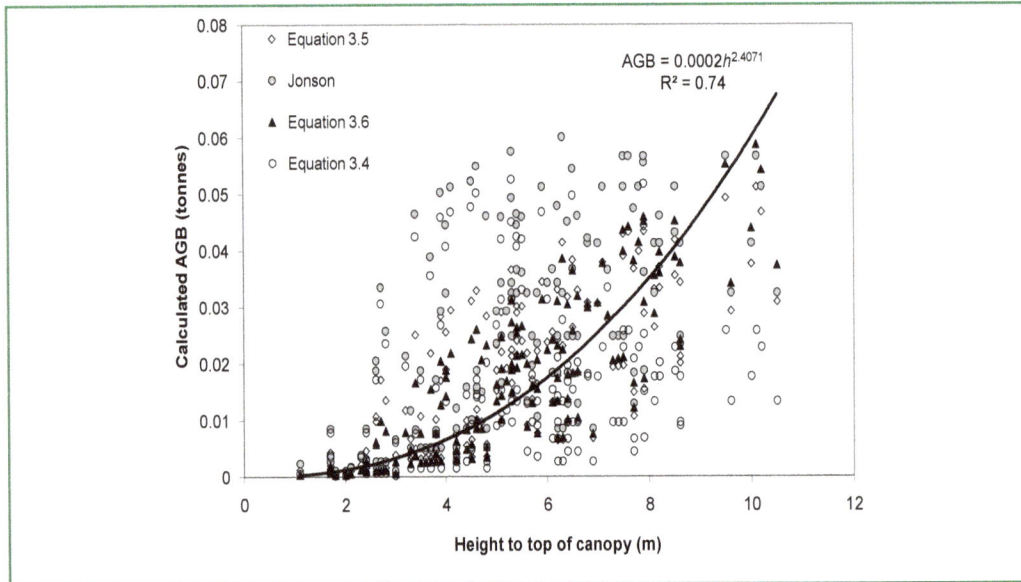

Figure 3.7 Relationship between measured canopy height and calculated AGB (tonnes per plant) for 114 small trees or shrubs having dbh <10 cm. The equation displayed on the graph (Equation 3.7) yields the best prediction of AGB (based on Equation 3.6) from height for these small trees and shrubs.

Below-ground biomass (BGB) or root biomass is usually estimated as a ratio of above-ground biomass. The most appropriate ratios to use for the GWW are based on site data for similar species and environmental conditions; estimates obtained for a range of Australian woodlands are summarised in Table 3.2. The value obtained from the generic equations of Jonson (n.d.) was in close agreement with the values given in Table 3.2 for Oakvale and Howard Springs. Therefore, we used the ratio of BGB:AGB obtained from the analysis of Jonson to estimate BGB at our field survey sites.

Table 3.2 Summary of estimated ratio of below-ground biomass to above-ground biomass for a range of Australian woodlands.

Species	Location	Rainfall (mm yr^{-1})	BGB:AGB	Source
E. populnea	Oakvale, NSW	367	0.58	Zerihun et al. (2006)
E. populnea	Roma, Qld	602	0.42	Zerihun et al. (2006)
E. miniata, E. tetradonta	Howard Springs, Humpty Doo, Wildlife Park, NT	1200–>1400	0.64	Chen et al. (2003)
Several eucalypts, see Table A3	Wheatbelt, WA	~300–500	0.60	Jonson

We had no field measurements of the three components of dead above-ground biomass (AGB_{dead}) for the GWW: standing deadwood (SDW), coarse woody debris (CWD) on the ground and litter (Lit). In a review of AGB_{dead} in Australian forests (Woldendorp and Keenan 2005), data from two woodland studies gave AGB_{dead} equivalent to 26 per cent of the total AGB_{total} (AGB_{living} + AGB_{dead}). For open forests, the percentage of AGB_{dead}:AGB_{total} (five studies) was 25 per cent. SDW was ~7 per cent of AGB_{total} in the TRAPS study in Queensland woodlands (Burrows et al. 2002). In contrast, the mean percentage of SDW:AGB_{total} reported by Woldendorp and Keenan (2005) was 15.6 per cent for the two woodland studies and 1.9 per cent for the five open-forest studies. Based on these studies, we assume that AGB_{dead}:AGB_{total} for woodlands within the GWW is 26 per cent.

Total biomass was calculated as the sum of AGB_{living}, BGB and AGB_{dead}. The site-based estimates of total biomass are given in Table 3.3. The sites having the greatest total biomass (79–198 t ha^{-1}) are eucalypt woodlands. In contrast, the total biomass of the mallee, heath and other shrubland sites ranges from 2 to 50 t ha^{-1}.

3.2 DATA LAYERS FOR SPATIAL EXTRAPOLATION

In this section, we investigate how a range of factors—for which we have spatial data— could influence biomass. Those factors that are shown to be important are consequently taken into account in the spatial modelling process by which we extrapolate from the 21 field sites to estimate the total biomass and carbon stock of the ~16 million ha of the GWW.

3.2.1 Vegetation mapping

The NVIS MVG classes (DEWHA 2005) compare reasonably well with our observations at our field sites (Table 3.4). In some cases, however (specifically sites 3, 4, 10, 11 and 21), we did not observe the upper tree or shrub layer, and in some cases (sites 1, 2 and 9) we observed a tree layer at sites classified by NVIS as mallee woodland and shrubland. One possible explanation for these disparities is that the vegetation has changed in the period since the NVIS mapping data were collected (which was at variable times before 2005 in different regions of Australia). As noted, the GWW vegetation has been subjected to several types of disturbance during the past century, including fire, timber cutting, mining and mineral exploration. The areas known to be affected by these disturbances are shown in Figure 2.14. Some parts of the GWW have also been impacted on by pastoralism.

Table 3.3 Estimated above-ground biomass, total biomass and corresponding aerial photographs of the 21 ANU field survey sites.

Site description	Site 1
	2005_01
	Eucalypt woodland
AGB estimate (t ha^{-1})	46
Total biomass (live+dead) (t ha^{-1})	102

Site description	Site 2
	2005_02
	Open eucalypt woodland
AGB estimate (t ha^{-1})	6
Total biomass (live+dead) (t ha^{-1})	14

Site description	Site 3
	2005_03
	Low shrubland
AGB estimate (t ha^{-1})	0.6
Total biomass (live+dead) (t ha^{-1})	1.3

Site description	Site 4
	2005_04
	Low shrubland
AGB estimate (t ha^{-1})	2
Total biomass (live+dead) (t ha^{-1})	5.2

Table 3.3 Continued

Site description	Site 5 2005_05 Eucalypt woodland
AGB estimate (t ha^{-1})	36
Total biomass (live+dead) (t ha^{-1})	79

Site description	Site 6 2005_06 Eucalypt woodland
AGB estimate (t ha^{-1})	48
Total biomass (live+dead) (t ha^{-1})	105

Site description	Site 7 2005_07 Eucalypt woodland
AGB estimate (t ha^{-1})	49
Total biomass (live+dead) (t ha^{-1})	108

Site description	Site 8 2005_08 Tall shrubland
AGB estimate (t ha^{-1})	2
Total biomass (live+dead) (t ha^{-1})	4.4

Table 3.3 Continued

Site description	Site 9 2005_09 Eucalypt woodland
AGB estimate (t ha^{-1})	32
Total biomass (live+dead) (t ha^{-1})	70

Site description	Site 10 2005_10 Tall shrubland
AGB estimate (t ha^{-1})	5
Total biomass (live+dead) (t ha^{-1})	11

Site description	Site 11 2006_01 Low eucalypt woodland
AGB estimate (t ha^{-1})	11
Total biomass (live+dead) (t ha^{-1})	24

Site description	Site 12 2006_02 Eucalypt woodland
AGB estimate (t ha^{-1})	58
Total biomass (live+dead) (t ha^{-1})	127

Table 3.3 Continued

Site description	Site 13 2006_03 Eucalypt woodland
AGB estimate (t ha^{-1})	47
Total biomass (live+dead) (t ha^{-1})	104

Site description	Site 14 2006_04 Eucalypt woodland
AGB estimate (t ha^{-1})	49
Total biomass (live+dead) (t ha^{-1})	108

Site description	Site 15 2006_05 Eucalypt woodland
AGB estimate (t ha^{-1})	38
Total biomass (live+dead) (t ha^{-1})	84

Site description	Site 16 2006_06 Tall acacia shrubland
AGB estimate (t ha^{-1})	4
Total biomass (live+dead) (t ha^{-1})	9.6

Table 3.3 Continued

Site description	Site 17 2006_07 Tall acacia shrubland
AGB estimate (t ha^{-1})	6
Total biomass (live+dead) (t ha^{-1})	13

Site description	Site 18 2007_01 Tall mallee shrub over low marlock, melaleuca understorey
AGB estimate (t ha^{-1})	23
Total biomass (live+dead) (t ha^{-1})	50

Site description	Site 19 2007_02 Eucalypt/melaleuca woodland
AGB estimate (t ha^{-1})	38
Total biomass (live+dead) (t ha^{-1})	84

Site description	Site 20 2007_03 Eucalypt woodland
AGB estimate (t ha^{-1})	90
Total biomass (live+dead) (t ha^{-1})	198

Table 3.3 Continued

Site description	Site 21 2007_04 Mallee/heath
AGB estimate (t ha^{-1})	2
Total biomass (live+dead) (t ha^{-1})	4.7

The aerial photographs are reproduced by permission of WA Land Information Authority, CL34/2010 (<www.landgate.wa.gov.au>).

Table 3.4 Comparison of vegetation description based on field observation and NVIS Major Vegetation Group (MVG) class.

Site no.	Survey ID	Field vegetation description	NVIS-MVG extant classification code	NVIS-MVG extant vegetation description
1	2005_01	Eucalypt woodland	14	Mallee woodlands and shrublands
2	2005_02	Open eucalypt woodland	14	Mallee woodlands and shrublands
3	2005_03	Low shrubland	14	Mallee woodlands and shrublands
4	2005_04	Low shrubland	14	Mallee woodlands and shrublands
5	2005_05	Eucalypt woodland	5	Eucalypt woodland
6	2005_06	Eucalypt woodland	5	Eucalypt woodland
7	2005_07	Eucalypt woodland	5	Eucalypt woodland
8	2005_08	Tall shrubland	17	Other shrubland
9	2005_09	Eucalypt woodland	14	Mallee woodlands and shrublands
10	2005_10	Tall shrubland	14	Mallee woodlands and shrublands
11	2006_01	Low eucalypt woodland	5	Eucalypt woodland
12	2006_02	Eucalypt woodland	5	Eucalypt woodland
13	2006_03	Eucalypt woodland	5	Eucalypt woodland
14	2006_04	Eucalypt woodland	5	Eucalypt woodland
15	2006_05	Eucalypt woodland	5	Eucalypt woodland
16	2006_06	Tall acacia shrubland	15	Low closed forest and tall shrubland
17	2006_07	Tall acacia shrubland	15	Low closed forest and tall shrubland
18	2007_01	Mallee shrubland	14	Mallee woodlands and shrublands
19	2007_02	Eucalypt woodland	5	Eucalypt woodland
20	2007_03	Eucalypt woodland	5	Eucalypt woodland
21	2007_04	Heath/mallee shrubland	5	Eucalypt woodland

Source: NVIS MVG DEWHA (2005).

3.2.2 Recent fire history

The occurrence of wildfire can be detected through the analysis of time-series of satellite imagery. We used remotely sensed data from two satellite-based sources—namely: Landsat images[7] at 30 m pixel resolution (1972, 1977, 1980, 1985, 1988, 1989, 1991, 1992, 1995, 1998, 2000 and 2002) and MODIS (or Moderate Resolution Imaging Spectroradiometer) images at 250 m pixel resolution beginning in 2000 (see Section 3.2.7 for details). While having a coarser pixel resolution, the MODIS data were processed to produce an essentially cloud-free time-series on a one-month time step (see Berry et al. 2007). Using these data, fire footprints over intensely burnt areas were highlighted through 'difference images', which show changes that have occurred between subsequent images. We used this analytical approach to map areas that were severely burnt by large landscape fires beginning circa 1972. The extent and frequency of these fires are shown in Figure 3.8.

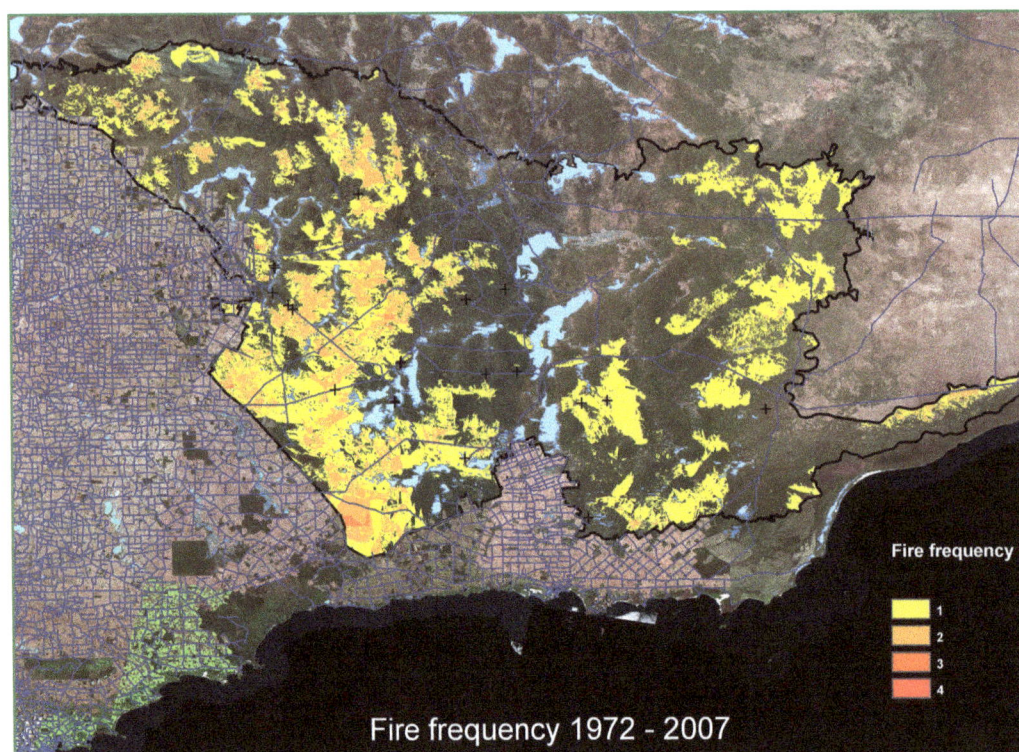

Fire frequency

1
2
3
4

Fire frequency 1972 - 2007

Figure 3.8 Area identified as being burnt by intense wildfires between 1972 and 2007. The base map is the Landsat 2005 layer. The fire frequency indicates the number of times an area within the GWW has been burnt by intense fires over the period. The base map is visible in areas within the GWW having a fire frequency of zero between 1972 and 2007. Fire frequencies are not shown for areas outside the GWW boundary. The salt lakes are coloured blue. Major roads are shown in blue and + signs indicate the location of ANU field survey sites.

Spatial data sources: see Table A1—Fire mapping, MODIS, WA Landcover, Roads. This figure incorporates Landsat 2005 data and road location data that are Copyright State of Western Australia 2007, and Landsat Continental Mosaic (AGO) data that are Copyright Commonwealth of Australia 2010.

We found that all five sites for which our vegetation structure observations differed from the NVIS MVG mapping were burnt since 1972, and mostly since 1990. This suggests that either: 1) at some time after the vegetation survey for mapping, fire has altered the vegetation structure; or 2) the vegetation map used to inform NVIS depicted the climax

---

7   See Table A1 for details.

vegetation instead of an intermediate successional state. Vegetation structure could change if fire caused the death of above-ground plant tissues so that subsequent regeneration is from below-ground tissues (lignotubers or roots) or from seed. The relative abundance of species can change due to differing rates of regeneration. If the fire return interval is too short then it is possible that lignotuber resources will be depleted and no re-sprouting will occur. Also, there will be insufficient time for production and accumulation of seed. This could lead to local loss of species and possibly a change in the species composition of the dominant canopy layer (for example, a eucalypt-dominated canopy could be replaced with an acacia-dominated canopy). Hopkins and Robinson (1981), in their study of the effects of a single fire in the GWW, concluded that the mallee-heath vegetation was a pyric disclimax vegetation following a fire in eucalypt woodland. Beard (1968:251), in his report on mapping of vegetation of the Lake Johnston and Boorabbin areas within the GWW, noted that the eucalypt trees that dominate sclerophyll woodland are killed by fires and regenerate from seed. Beard also states that '[m]allee is subject to frequent fires which destroy the top growth, regeneration taking place from coppice. It is not clear to what extent the mallee habit is genetically controlled or due to fire.' Beard notes that 'very many' mallee species can be found in the form of moderately sized or small trees. In his description of the 'scrub heath' vegetation formation, Beard (1968:249) observed that this vegetation is 'burnt so frequently that a mature structure has little chance to develop'. The observations of Hopkins and Robinson, and Beard, suggest that fire has played a pivotal role in modifying the vegetation structure in the GWW. Observations during field surveys between 2005 and 2007 by ANU researchers support those of Hopkins and Robinson (1981) and Beard (1968) (Figures 3.9 and 3.10).

Figure 3.9b

Figure 3.9d

Figure 3.9a

Figure 3.9c

Figures 3.9 Early stages of regeneration of eucalypts after fire. Fire has killed the tree stratum. For the purpose of carbon accounting, the re-growth vegetation is classified as 'low closed forests/tall closed shrublands' (NVIS MVG 15).

Photos: S. Berry.

Figure 3.10b

Figure 3.10a

Figures 3.10 This surviving fire-damaged tree (a) testifies to the pre-fire woodland state of the vegetation (NVIS MVG 5), which would currently be classified as mallee woodland or shrubland (MVG 14) or low closed forest/tall closed shrubland (NVIS MVG 15). If fire is excluded for several decades, succession of the vegetation in (a) will lead to sclerophyll woodland (NVIS MVG 5) with a similar structure to that shown in (b).

If the vegetation structure mapping in the GWW includes pyric successional vegetation (such as that shown in Figure 3.9), we would expect congruence between boundaries of vegetation mapping units and fire footprints in addition to congruence between the vegetation and surficial geology boundaries[8]. We present evidence for congruence between fire footprints and vegetation boundaries in Figures 3.11 and 3.12. The mapped (MVG 5) eucalypt woodland boundaries border many footprints of fires in the period 1972–90 (Figure 3.11a). Post-1990 fire footprints occur over large areas that are mapped as eucalypt woodland (Figure 3.11b), as would be expected if the vegetation map layer was derived from pre-1990 information. The presence of fire footprints over vegetation mapped as woodland (Figure 3.11b), along with our field observations (Figures 3.9 and 3.10), demonstrates that the eucalypt woodland has been extensively burned in the past two decades. This can be contrasted with the observation of Beard (1968:232) that woodlands, when he observed them, burned to only a minor extent.

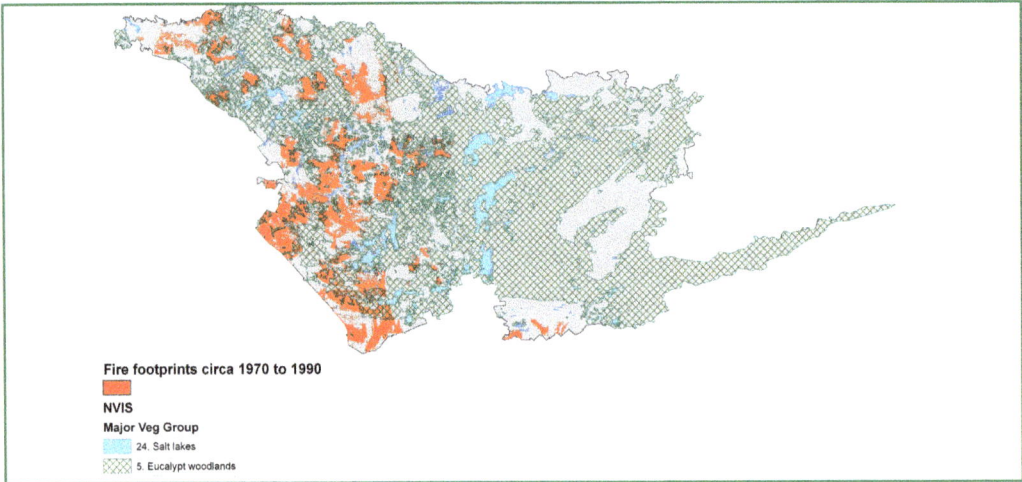

Figure 3.11a

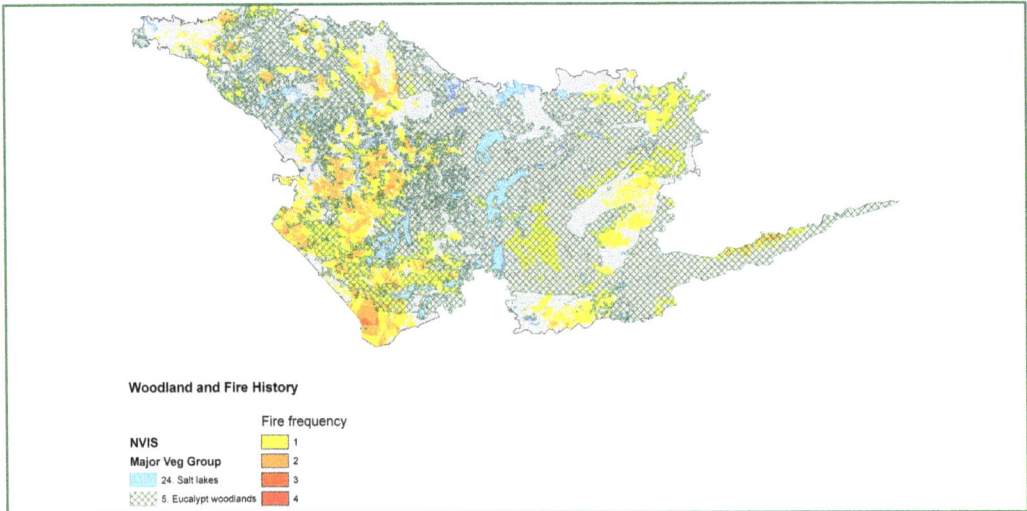

Figure 3.11b

Figures 3.11 Location of fire footprints identified from (a) Landsat satellite imagery for the period 1972–90, and (b) both Landsat and MODIS imagery for the period 1972–2007. The distribution of eucalypt woodland (NVIS MVG 5) is shown as an overlay.

These figures incorporate NVIS Version 3.0 data and Landsat Continental Mosaic (AGO) data that are Copyright Commonwealth of Australia 2005 and 2010.

---

8   As noted in Section 2.3 and Table A2, however, the derivation of the vegetation and surface geology layers for some map sheets relied on interpretation from aerial photographs, the surficial geology boundaries being inferred from vegetation cover.

The example presented in Figure 3.12—an area that was disturbed by timber cutting between 50 and 100 years ago—also provides some evidence to support the proposition that eucalypt woodland can be replaced by shrubland dominated by single-stemmed eucalypts (marlock), mallee or acacia in the decades after intense fire. Eucalypt woodlands (MVG5) are mapped as occurring over a wide range of surface geology classes, with geology having considerable influence on the species represented (Table A3). In Figure 3.12, low closed forest/tall closed shrubland (MVG 15) vegetation mapping coincides with the mapping of the Czs (Cainozoic sands) geology class, while adjoining woodland vegetation is mapped on the Czg (Cainozoic gravel) geology. This correspondence between vegetation classes and geology classes is, however, not consistent across all of the map sheets. On some sheets, eucalypt woodland is mapped over Czs.

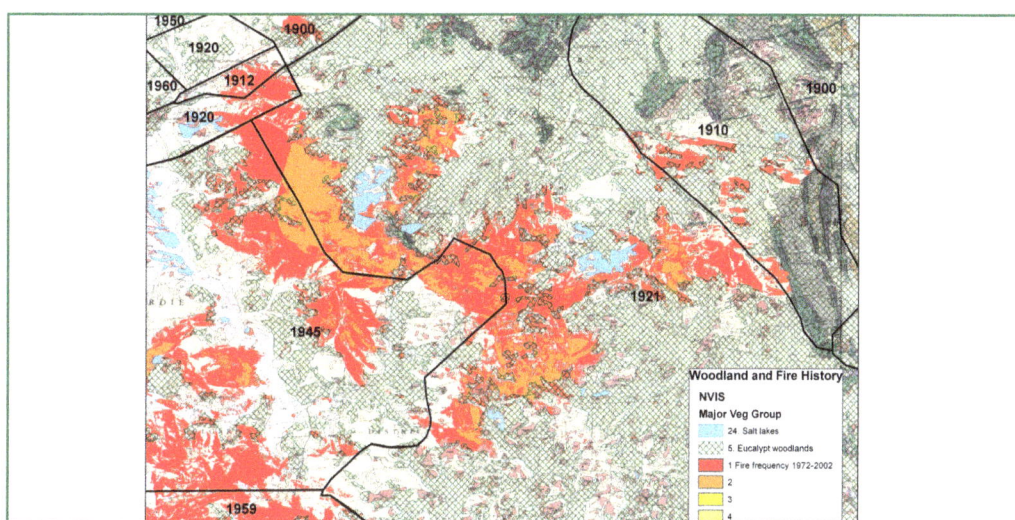

Figure 12a

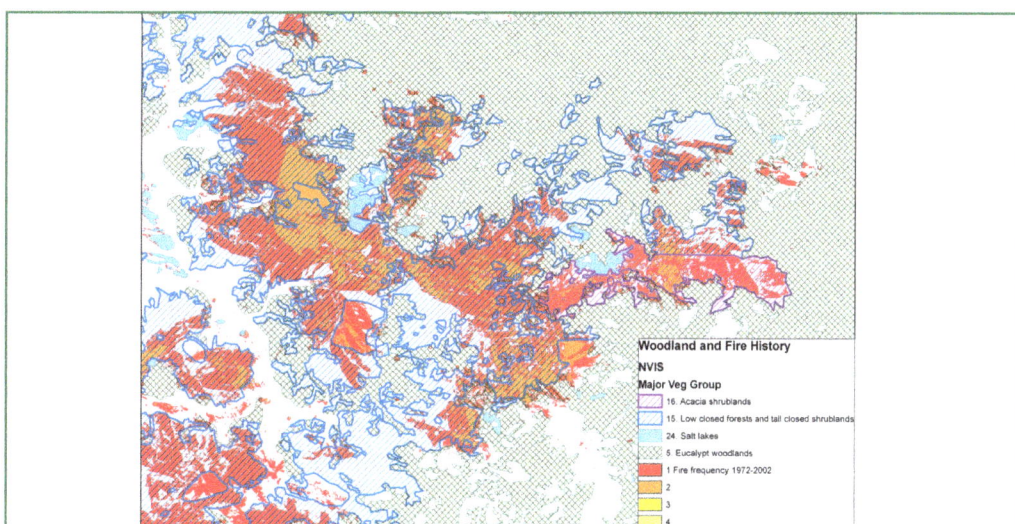

Figure 3.12b

Figures 12 Congruence between fire footprint and eucalypt woodland boundaries in part of the GWW. (a) The vegetation and fire layers are superimposed on the 1:250 000 surface geology map for Boorabbin. The entire area is mapped as being cut over for mining timber and/or fuel wood. The year in which timber cutting began within each section (outlined in black) is shown. (b) The extant vegetation over the fire footprint is mapped as shrubland (NVIS MVG 15 and 16).

Data source: see Table A1—Fire Mapping, Map Sheet, Geology, Timber cutting, MVG. This figure incorporates geology data that are Copyright State of Western Australia 2007, and NVIS Version 3.0 data and Landsat Continental Mosaic (AGO) data that are Copyright Commonwealth of Australia 2005 and 2010.

Given the above, and as in the semi-arid GWW, the eucalypt seedlings and saplings making up the post-fire regeneration of woodland form a tall shrubland (MVG 15; Figure 3.9) for several decades following fire, we have 'updated' the mapped distribution of extant eucalypt woodland to account for the changes wrought by intense fires in the past 35 years by reclassifying the 1.46 million ha of MVG 5 (eucalypt forests and woodlands) vegetation that has been severely burnt by fires between 1972 and 2007 into MVG 15 (low closed forests and tall closed shrublands) (Figure 3.13).

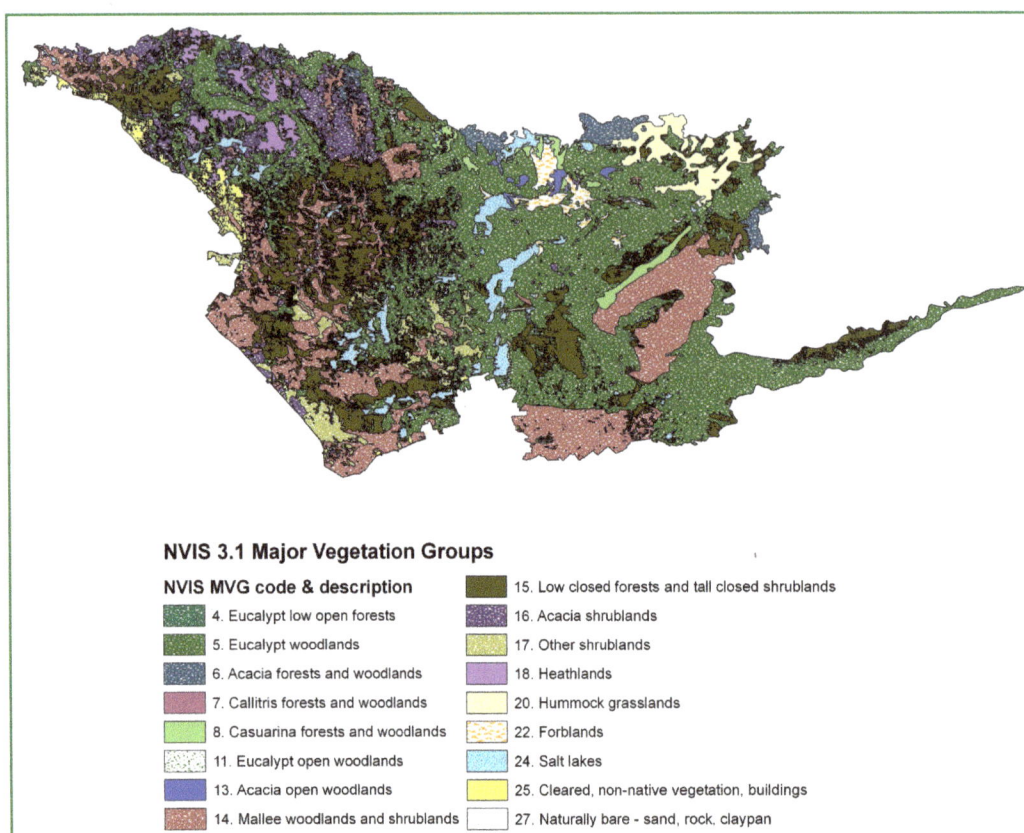

NVIS 3.1 Major Vegetation Groups

NVIS MVG code & description

4. Eucalypt low open forests		15. Low closed forests and tall closed shrublands	
5. Eucalypt woodlands		16. Acacia shrublands	
6. Acacia forests and woodlands		17. Other shrublands	
7. Callitris forests and woodlands		18. Heathlands	
8. Casuarina forests and woodlands		20. Hummock grasslands	
11. Eucalypt open woodlands		22. Forblands	
13. Acacia open woodlands		24. Salt lakes	
14. Mallee woodlands and shrublands		25. Cleared, non-native vegetation, buildings	
		27. Naturally bare - sand, rock, claypan	

Figure 3.13 This map shows the likely current distribution of vegetation structural formations (NVIS MVGs) in the GWW in 2008. It was made by reclassifying the 1.46 million ha of NVIS MVGs 4 and 5 (eucalypt forests and woodlands) vegetation that have been severely burnt by fires between 1972 and 2007 into MVG 15 (low closed forests and tall closed shrublands) to better describe the structure of the post-fire regeneration of seedlings and root sprouts.

Data source: see Table A1—MVG. This figure incorporates NVIS Version 3.0 data that are Copyright Commonwealth of Australia 2005.

The mapping of fire footprints (Figure 3.11) is restricted to fires after 1972. Footprints of fires predating that time are, however, also evident in the Landsat satellite imagery. If we assume that the NVIS MVG classes that include Acacia and mallee woodlands and shrublands (MVGs 13, 14, 15 and 16) arise following intense fire in eucalypt woodlands and forests in the GWW, we can predict the extent of the latter in a 'no-fire' condition, and this is shown in Figure 3.14. In this scenario, the eucalypt woodlands occupy 13 million ha of the GWW—double the current extent. This estimate is likely greater than the actual extent of eucalypt woodland (MVG 5) structure pre-1750, as our scenario assumes no fire and no geological differences underlying the distributions of eucalypt woodlands and mallee formations.

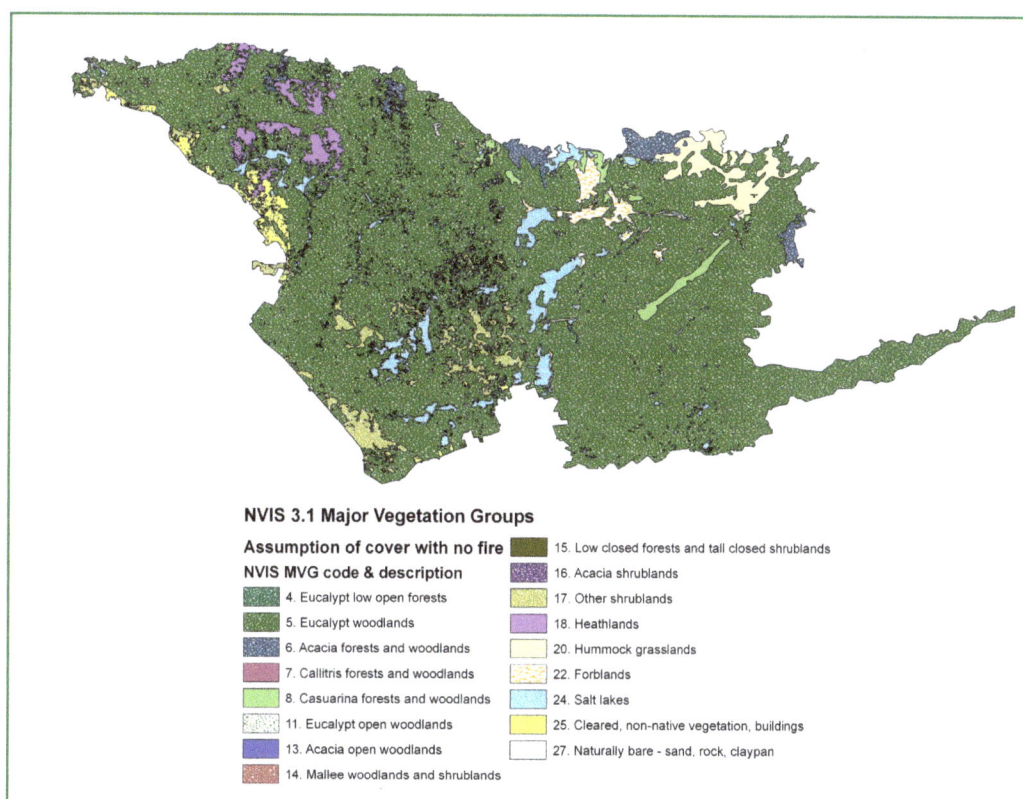

NVIS 3.1 Major Vegetation Groups

Assumption of cover with no fire

NVIS MVG code & description

- 4. Eucalypt low open forests
- 5. Eucalypt woodlands
- 6. Acacia forests and woodlands
- 7. Callitris forests and woodlands
- 8. Casuarina forests and woodlands
- 11. Eucalypt open woodlands
- 13. Acacia open woodlands
- 14. Mallee woodlands and shrublands

- 15. Low closed forests and tall closed shrublands
- 16. Acacia shrublands
- 17. Other shrublands
- 18. Heathlands
- 20. Hummock grasslands
- 22. Forblands
- 24. Salt lakes
- 25. Cleared, non-native vegetation, buildings
- 27. Naturally bare - sand, rock, claypan

Figure 3.14 Predicted extent of eucalypt forest and woodland and other vegetation groups assuming fire was absent from the GWW for many decades to centuries and NVIS MVGs 13–16 occur at present in locations that were previously (that is, before a past fire event) eucalypt woodland (MVG 5). We subsequently refer to this hypothetical vegetation cover as the 'no-disturbance scenario'.

Data source: see Table A1—MVG. This figure incorporates NVIS Version 3.0 data that are Copyright Commonwealth of Australia 2005.

3.2.3 Timber cutting

The approximate location of timber-cutting activities in the GWW during the period 1900–75 (digitised from the map shown in Figure 2.12) is shown superimposed over the NVIS MVG layer in Figure 3.15. Timber cutting in this period affected approximately 4.4 million ha of vegetation in the GWW (Table 3.5). Historical records indicate that timber cutting in the GWW utilised mostly the salmon gum and gimlet woodlands (see Section 2.5). Following timber cutting, several species of eucalypt are able to regenerate from coppice (the authors have observed this at their field survey sites). The coppice growth arises from the top of the stump of the cut tree so that the original monopodial tree retains a monopodial base, which supports multiple stem sprouts. In contrast, mallee sprouts arise from the base of the stem to form a multi-stemmed shrub or tree.

According to our map (Figure 3.15) and Table 3.5, eucalypt woodlands (NVIS extant MVG 5 and 11) make up only 2.5 million ha of the timber-cutting area. However, 1.1 million ha of the area cut over for timber has a mapped vegetation cover class of mallee woodland and shrubland (MVG 14), tall closed shrubland (MVG 15) or acacia shrubland (MVG 16) (Table 3.5 and Figure 3.15). Half of the 1.1 million ha of MVGs 14, 15 and 16 has subsequently been burned by intense fires. In Section 3.2.2, we proposed that MVGs 14, 15 and 16 could represent pyric successional vegetation. If fire has altered the vegetation structure in this way following timber cutting then prior to timber cutting the vegetation cover could have been similar to the modelled vegetation cover shown in Figure 3.14. We show this modelled vegetation cover, beneath the timber-cutting overlay, in Figure 3.16. Under the 'no-disturbance' scenario, 3.6 million ha of eucalypt woodland would have been available for timber cutting circa 1900.

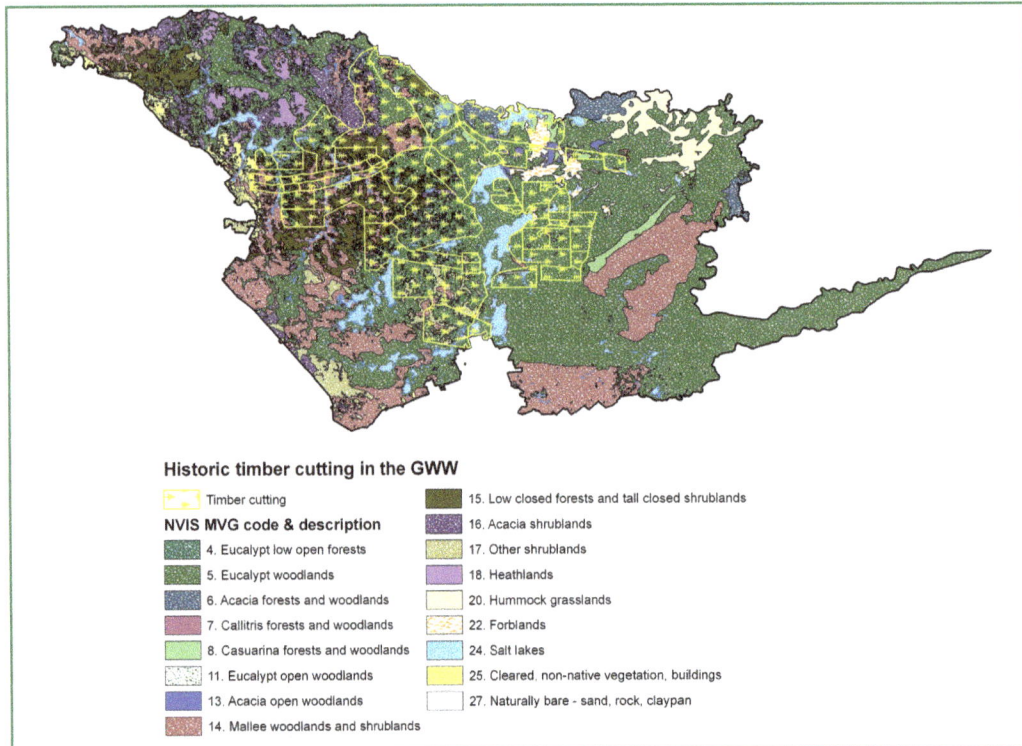

Historic timber cutting in the GWW

Timber cutting	15. Low closed forests and tall closed shrublands
NVIS MVG code & description	16. Acacia shrublands
4. Eucalypt low open forests	17. Other shrublands
5. Eucalypt woodlands	18. Heathlands
6. Acacia forests and woodlands	20. Hummock grasslands
7. Callitris forests and woodlands	22. Forblands
8. Casuarina forests and woodlands	24. Salt lakes
11. Eucalypt open woodlands	25. Cleared, non-native vegetation, buildings
13. Acacia open woodlands	27. Naturally bare - sand, rock, claypan
14. Mallee woodlands and shrublands	

Figure 3.15 Extent of timber cutting in the GWW between 1900 and 1975 is shown as an overlay on the NVIS MVG extant vegetation layer.

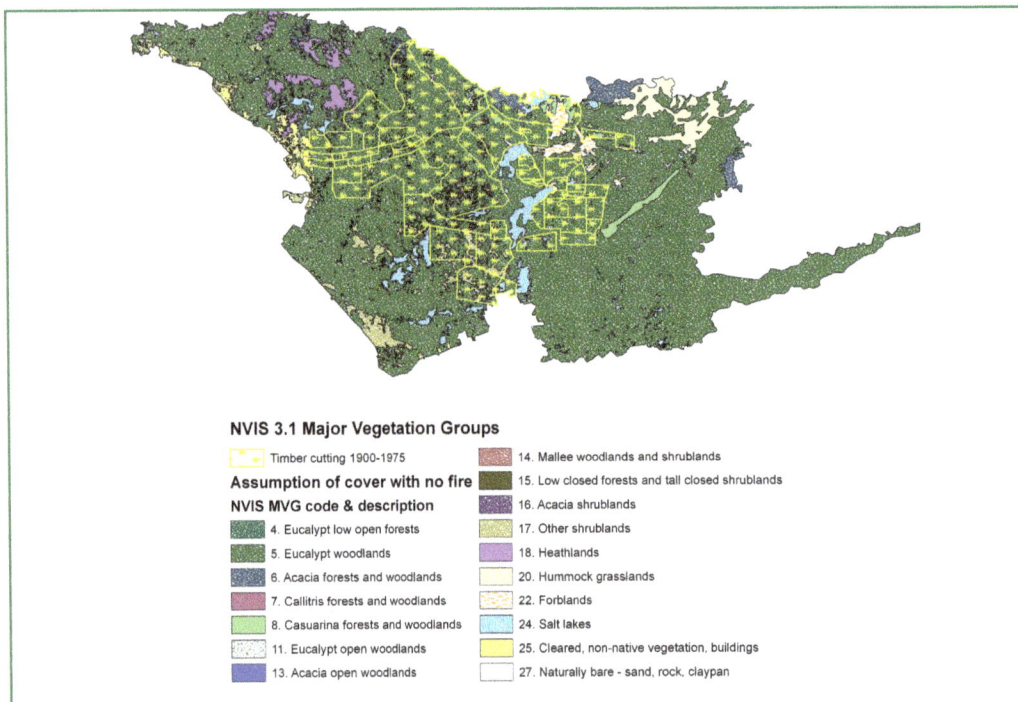

NVIS 3.1 Major Vegetation Groups

Timber cutting 1900-1975	14. Mallee woodlands and shrublands
Assumption of cover with no fire	15. Low closed forests and tall closed shrublands
NVIS MVG code & description	16. Acacia shrublands
4. Eucalypt low open forests	17. Other shrublands
5. Eucalypt woodlands	18. Heathlands
6. Acacia forests and woodlands	20. Hummock grasslands
7. Callitris forests and woodlands	22. Forblands
8. Casuarina forests and woodlands	24. Salt lakes
11. Eucalypt open woodlands	25. Cleared, non-native vegetation, buildings
13. Acacia open woodlands	27. Naturally bare - sand, rock, claypan

Figure 3.16 Extent of timber cutting in the GWW between 1900 and 1975 is shown as an overlay on the 'no-disturbance scenario' modelled vegetation layer (see Figure 3.14).

Data source: see Table A1—Timber cutting, MVG. These figures incorporate NVIS Version 3.0 data that are Copyright Commonwealth of Australia 2005.

Table 3.5 Area within each NVIS MVG class for extant (Figure 3.15) and 'no-disturbance' (Figure 3.16) scenario vegetation, within the area mapped as affected by timber cutting between 1900 and 1975. The area of each vegetation type affected by both timber cutting and fire (1972–2007) is also shown.

MVG class	Area affected by timber cutting (ha)	Area affected by timber cutting and fire (ha)	Area affected by timber cutting (ha)
	NVIS extant vegetation	NVIS extant vegetation	Hypothetical 'no-disturbance' scenario
4. Eucalypt low open forest	0	0	0
5. Eucalypt woodlands	2 486 560	167 898	3 614 603
6. Acacia forests and woodlands	94 515	833	94 515
7. Callitris forests and woodlands	277	7	277
8. Casuarina forests and woodlands	117 155	0	117 155
11. Eucalypt open woodlands	5 746	812	5 746
13. Acacia open woodlands	21 121	0	0
14. Mallee woodlands and shrublands	411 304	117 915	0
15. Low closed forests and tall closed shrublands	563 713	341 064	0
16. Acacia shrublands	131 905	63 790	0
17. Other shrublands	204 881	46 961	204 881
18. Heathlands	4 734	285	4 734
20. Hummock grasslands	4 418	3 601	4 418
22. Forblands	74 505	1 268	74 505
24. Salt lakes	119 874	5 098	119 874
25. Cleared, non-native vegetation	11 816	3 607	11 816
27. Naturally bare—sand, rock, claypan	115 472	25 721	115 472
Total	**4 367 997**	**778 859**	**4 367 997**

3.2.4 Mining and mineral exploration

Most of the mining and mineral exploration in the GWW is located within the boundaries of the greenstone lithological unit (Figure 2.6). The area impacted by open-cut mining could be relatively small. Mineral exploration has, however, more profound impacts on the vegetation including the use of vehicles off-road to access remote country, construction of access tracks, excavations to expose deeper soil horizons, construction of seismic lines and the cutting down and removal of wood for fuel, mine-props, and so on (see Plates 1–5). A map showing the location of the major lithological units and the land tenure type is presented in Figure 3.17. A large portion of the granite-greenstone lithological unit is under leasehold tenure and is used for mining and mineral exploration.

Plate 1 Professor Brendan Mackey, ANU, standing in a typical trench created during mineral exploration, GWW.

Plate 2 Aerial photograph of an open-cut mine site, GWW.

Reproduced by permission of WA Land Information Authority, CL34/2010 (<www.landgate.wa.gov.au>).

Plate 3 Seismic tracks near sites 13 and 14, GWW.

Reproduced by permission of WA Land Information Authority, CL34/2010 (<www.landgate.wa.gov.au>).

Plate 4 Aerial photograph showing the environmental footprint of a nickel mine, GWW.

Reproduced by permission of WA Land Information Authority, CL34/2010 (<www.landgate.wa.gov.au>).

Plate 5 Stump and felled tree, Site 14. Several trees had been felled at this remote site located on the greenstone belt.

3.2.5 Land tenure

Four types of land tenure predominate in the GWW: vacant crown land (55 per cent), nature conservation reserve (10 per cent), Aboriginal leasehold (1.5 per cent) and other leasehold (20 per cent) (see Figure 3.17). The area of coverage of each land tenure type, the land tenure of the extant eucalypt woodland and the land tenure of the mapped fire footprint (Figure 3.8) are given in Table 3.6. Thirty-two per cent of the extant woodland (as mapped in Figure 3.13) occurs on leasehold land, while 55 per cent is on vacant crown land and 9 per cent is within nature conservation reserves. Some land tenure types appear to be more prone to fire and approximately 36 per cent of the vacant crown land, nature conservation reserve and Aboriginal reserve classes bear a fire footprint (1972–2007 fire events) in contrast with 6 per cent of leasehold land. The lesser fire footprint on leasehold land could be a consequence of: 1) less fuel due to land clearing and grazing; 2) fire protection and suppression to protect economic assets; and/or 3) fewer ignition events due to restricted access.

For the purpose of carbon accounting, it was necessary to determine if the vegetation structure (and biomass) has been modified by land clearing on different land tenures in addition to the mapped timber cutting (area of timber cutting shown in relation to land tenure in Figure 3.17). As all of our field study sites were located on vacant crown land or within nature conservation reserves, we needed to utilise other data sources to assess the vegetation cover and biomass of leasehold land. To do this, we utilised remotely sensed imagery, as discussed in Section 3.3 below. Nonetheless, it is evident from Figure 3.17 that most of the land under leasehold tenure overlies granite-greenstone lithology (thus is likely to have been impacted on by mineral exploration) and/or has been subjected to timber cutting.

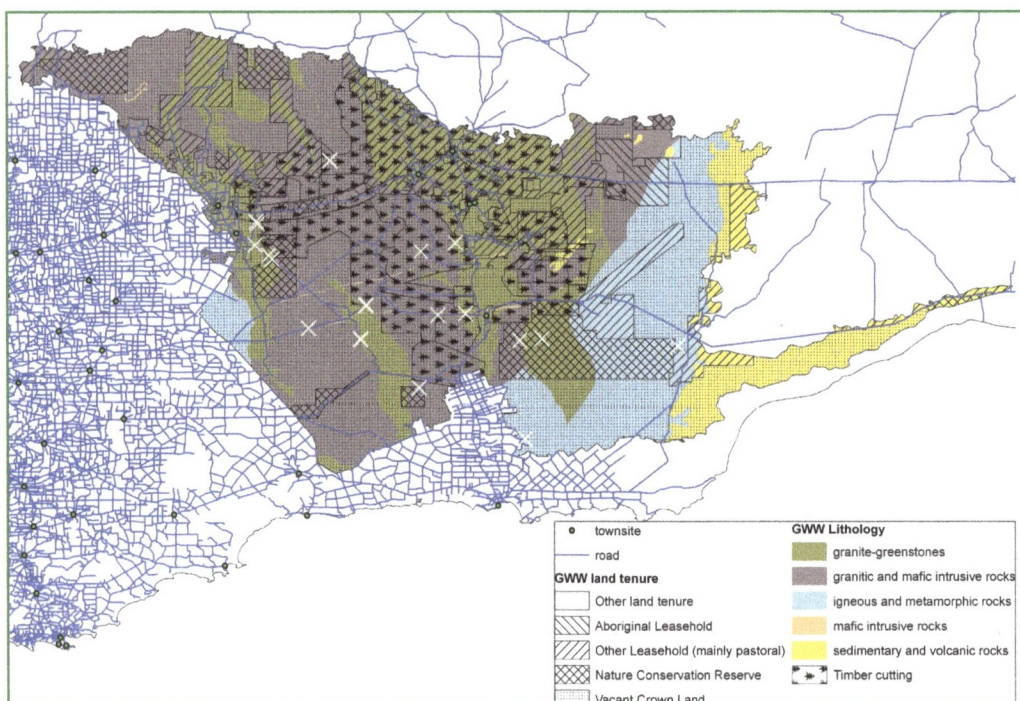

Figure 3.17 Overlay showing location of timber cutting superimposed on land tenure and lithology layers (Figure 2.6).

Data source: see Table A1—Timber cutting, Land Tenure, Lithology, Roads, Towns. This figure incorporates geology and road location data that are Copyright State of Western Australia 2007, and Australian land tenure (1993) and population centre data that are Copyright Commonwealth of Australia 1993 and 1998.

Table 3.6 Areal extent of land tenure types within the entire GWW, within the area mapped as woodland and within the area mapped as having a fire footprint between 1972 and 2007 (see Figure 3.8).

Land tenure type	Total area (ha × 1000)	Woodland area *(ha × 1000)	Total area burnt 1972–2007 (ha × 1000)
1. Aboriginal reserve	108	29	39
2. Aboriginal leasehold	252	170	42
3. Forestry reserve	120	88	3
4. Other freehold	249	16	13
5. Other leasehold	3 513	2 072	204
6. Mixed	9	9	0
7. Nature conservation reserve	1 651	605	590
8. Other	160	104	3
9. Vacant crown land	9 859	3 809	3 493
10. Water reserve	44	11	4

3.2.6 Overview of disturbance impacts

Our analyses of the available spatial data in this chapter have revealed that most of the GWW has experienced disturbance from the mining industry through minerals exploration and timber cutting, from clearing or thinning of the vegetation for cropping and grazing or from intense fire (Figure 3.18). All of these disturbances must be taken into account when extrapolating from field-based measurements of the biomass carbon stock to the entire GWW.

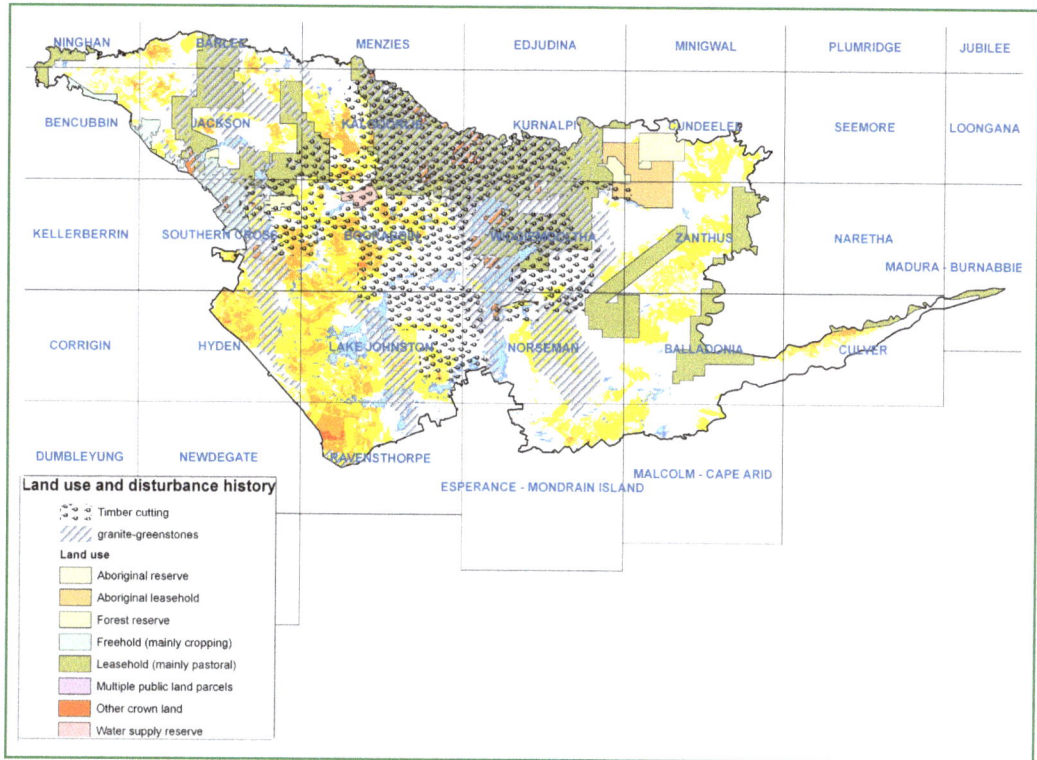

Figure 3.18 This map shows the extent of direct (timber cutting, minerals exploration and land-use change since 1900) and indirect (fire between 1972 and 2007) disturbances in the GWW. The legend for the fire footprints is shown in Figure 3.8.

Data source: see Table A1—Timber cutting, Land Tenure, Lithology, Map sheet. This figure incorporates geology and map sheet Index250 data that are Copyright State of Western Australia 2007, and Australian land tenure (1993) data that are Copyright Commonwealth of Australia 1993.

3.2.7 Relationship between land use/disturbance and AGB estimates for field survey sites

There is evidence that fire has had a larger effect than timber cutting on the above-ground biomass (AGB) at the ANU field survey sites (see Figure 3.19). The AGB of all field survey sites that have been burnt by intensive fire since 1972 is low. AGB of the ANU woodland sites that have been cut over is, however, more comparable with that of most of the other woodland sites. Further field surveys are needed to identify whether these relationships hold more broadly across the GWW, along with data on AGB values before and post-timber cutting. The field survey sites having the highest AGB occur in woodland where there has been no timber cutting, no fire since 1972 and no mining activity (that is, not on granite-greenstone lithology).

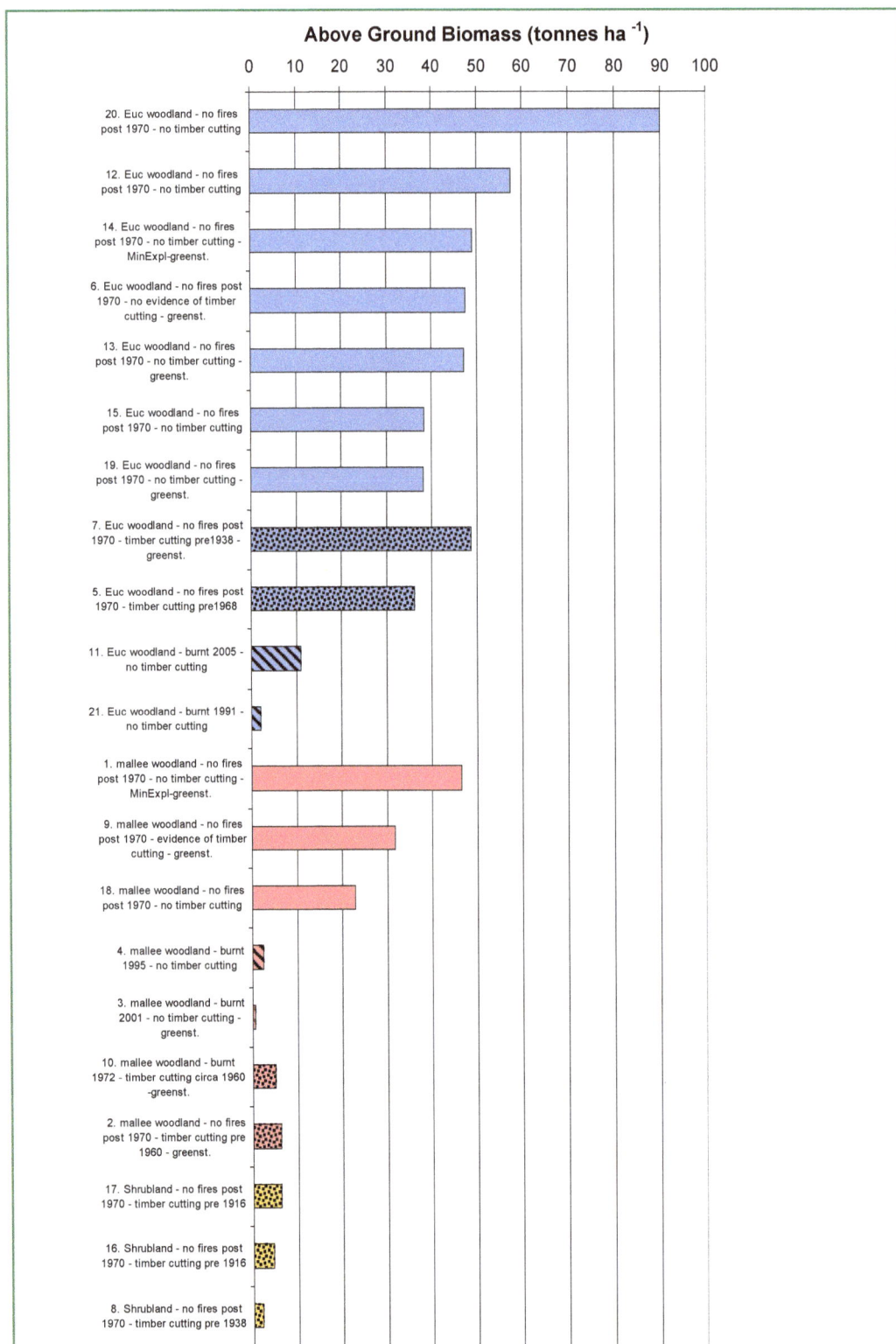

Above Ground Biomass (tonnes ha^{-1})

Figure 3.19 Relationship between NVIS MVG vegetation class, fire history (post ~1970), timber-cutting history (1900–75), surface lithology and AGB of ANU field survey sites (numbered 1–21). Fire, timber-cutting histories and lithology are inferred from the spatial data layers and field observation. The lithology is indicated only for sites occurring on the granite-greenstone formation, indicated here by 'greenst'. Our field notes record impacts of prospecting (vehicle tracks, excavations, and so on) on or close to many sites located on the granite-greenstone lithology. Although our timber-cutting spatial data layer indicated that Site 6 had been cut over, we could find no evidence of this activity having occurred. There were large trees present and no cut stumps or coppice growth.

3.3 EXTRAPOLATION OF CARBON STOCKS FROM SITE DATA

In the preceding section, we investigated the extent of disturbance by fire, timber cutting and mineral exploration in the GWW and identified those field survey sites that had been impacted on by these factors. Variability in the availability of water and nutrients across the GWW also affects foliage cover and must be taken into account when extrapolating from field survey sites to other locations.

3.3.1 Remote sensing of foliage cover

Spatial information about the green foliage cover of woody evergreen vegetation can be obtained through the analysis of time-series of satellite imagery (Berry and Roderick 2002; Roderick et al. 1999). We derived a spatial layer depicting the fraction of photosynthetically active radiation intercepted by the evergreen vegetation canopy (F_E) from MODIS 16-Day L3 Global 250m (MOD13Q1) satellite imagery for the period from July 2000 to July 2005 (Berry et al. 2007)[9].

The first step in the calculation of F_E involved the calculation of monthly values of the normalised difference vegetation index (NDVI). We then calculated the fraction of photosynthetically active radiation intercepted by the sunlit vegetation canopy (F_{PAR}) for all months in the time-series using the following equation:

$$F_{PAR}^* = 1.118NDVI^* - 0.168 \text{ in which } NDVI^* \geq 0.15$$

$$\text{and } F_{PAR}^* = 0 \text{ in which } NDVI^* < 0.15$$

in which '*' indicates a value for a specific month and year. We assumed that when NDVI <0.15 the surface cover is bare soil or rock. In any month, F_{PAR} will comprise a persistent evergreen component (F_E) due mostly to woody vegetation (Roderick et al. 1999) and a fluctuating seasonal component (F_R)—the green-flush following rain that is due mostly to herbaceous vegetation. To separate these components, we followed the general method of Donohue et al. (2009), but applied a 12-month moving minimum window and a 15-month smoothing function. We then determined the 12-month period from January 2001 to December 2004 for which F_E was at its maximum value and calculated a 12-month average F_E. This approach was taken in order to reduce the 'noise' from short-term defoliation events. The resulting spatial layer of average F_E is shown in Figure 3.20.

Figure 3.20 Fraction of photosynthetically active radiation intercepted by the evergreen component of the vegetation canopy (F_E). Units: x 100. Land cover types having low F_E include salt lakes, rock outcrops, agricultural land and vegetation canopies that have been thinned by fire or land clearing.

---

9   For data source for MODIS data, see Table A1—MODIS.

3.3.2 Relationship between foliage cover and AGB

The variable F_E provides information about the density of green foliage cover but not vegetation height, hence a direct relationship to biomass cannot be assumed. Over large environmental gradients and wide-ranging vegetation types, this relationship is a reasonable assumption because the environmental factors that influence canopy cover also influence height and biomass. Our field data for the GWW demonstrated, however, the lack of a general relationship between F_E and biomass: F_E of shrubland sites ranged from 0.04 to 0.37 and eucalypt woodland sites from 0.22 to 0.32. Thus, there were shrubland sites with canopies as densely green as some woodland sites. Within these two broad vegetation groups and woodland disturbance categories, however, relationships between F_E and AGB were evident. We found that although many of our woodland sites had very similar F_E values, sites that had been disturbed by timber cutting or mineral exploration had lower AGB than undisturbed woodland sites. This finding indicates that the canopy (comprising leaves supported by small branches) at disturbed sites has had sufficient time to recover to the pre-disturbance condition, while the mass (and volume) of woody stems that support the canopy remain below the pre-disturbance condition. We derived four equations to predict AGB (t ha^{-1}) from F_E (expressed as a percentage) in the GWW (see Figure 3.21). Three of these equations predict the AGB of eucalypt woodlands (NVIS MVG 4 and 5) for the three scenarios

i) no timber cutting, no mineral exploration disturbance (Equation 3.8)

Equation 3.8

$AGB = 0.0841F_E{}^2$

ii) mineral exploration, no timber cutting (Equation 3.9)

Equation 3.9

$AGB = 0.0597F_E{}^2$

iii) timber cutting (Equation 3.10)

Equation 3.10

$AGB = 0.0441F_E{}^2$

The fourth equation predicts AGB for all other woody vegetation classes of heath, mallee, marlock and shrub (Equation 3.11).

Equation 3.11

$AGB = 0.06F_E{}^{1.642}$

All of the equations above have been forced through (0, 0) although the line of best fit for scenario (i) would pass through a point where F_E exceeds 0 when AGB equals 0—a condition where a small woody vegetation canopy exists with no mass. In the case of scenarios (ii) and (iii), the line of best fit is plotted through a group of three points. The scatter between points is likely due to the variable impacts of mineral exploration and timber cutting at these sites. These activities have occurred over a long period (75 years in the case of timber cutting). Both the time period since disturbance and the intensity of disturbance affect AGB. It is not possible to develop more specific equations without the availability of detailed information describing these past disturbances and a greater number of field survey sites within each disturbance class. For our field sites, the effect of disturbance on biomass was greater than the influence of other environmental variables. More field data would be required to calibrate the effect of other environmental variables within disturbance classes.

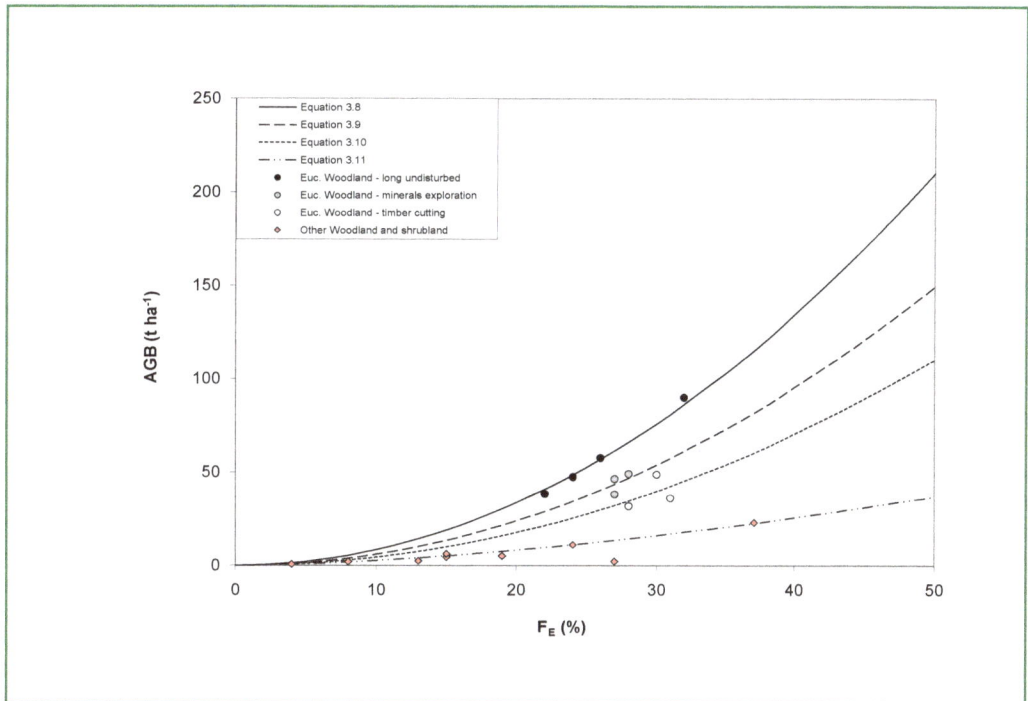

Figure 3.21a

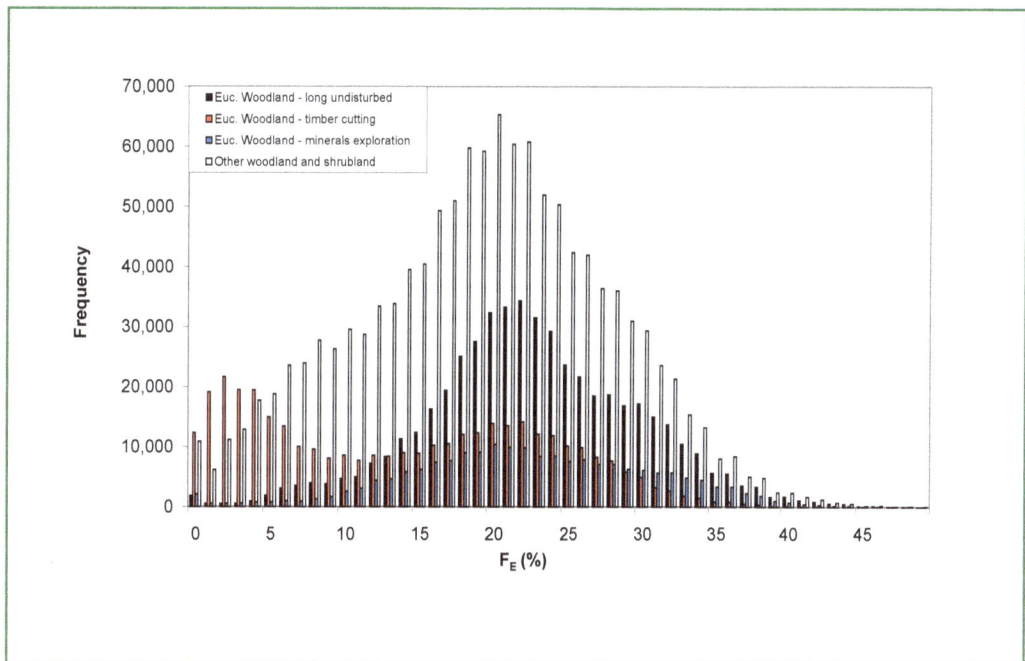

Figure 3.21b

Figures 3.21 Relationships between above-ground biomass (AGB, t ha^{-1}) and F$_E$ for the 21 field sites. F$_E$ values for the field sites were extracted from the spatial data layer shown in Figure 3.20. Equations 3.8–3.11 (see 'Vegetation—Remote Sensing' section) are also plotted. One equation is suitable for predicting the AGB of all vegetation structures other than eucalypt woodland. (b) The frequency histogram shows the distribution of 250 m × 250 m grid cells within F$_E$ classes for each of the AGB scenarios. The open bars indicate the maximum value of F$_E$ of field sites within the 'Other woodland and shrubland' class. The black, blue and red bars indicate the maximum values of F$_E$ of the eucalypt woodland classes.

We tested the capacity of Equations 3.8–3.11 to predict AGB by plotting the predicted values against those derived from measurements for our 21 field survey sites (Figure 3.22). Overall, there was good agreement between the actual and predicted values ($AGB_{predicted} = 0.99\ AGB_{actual}$; $R^2 = 0.93$).

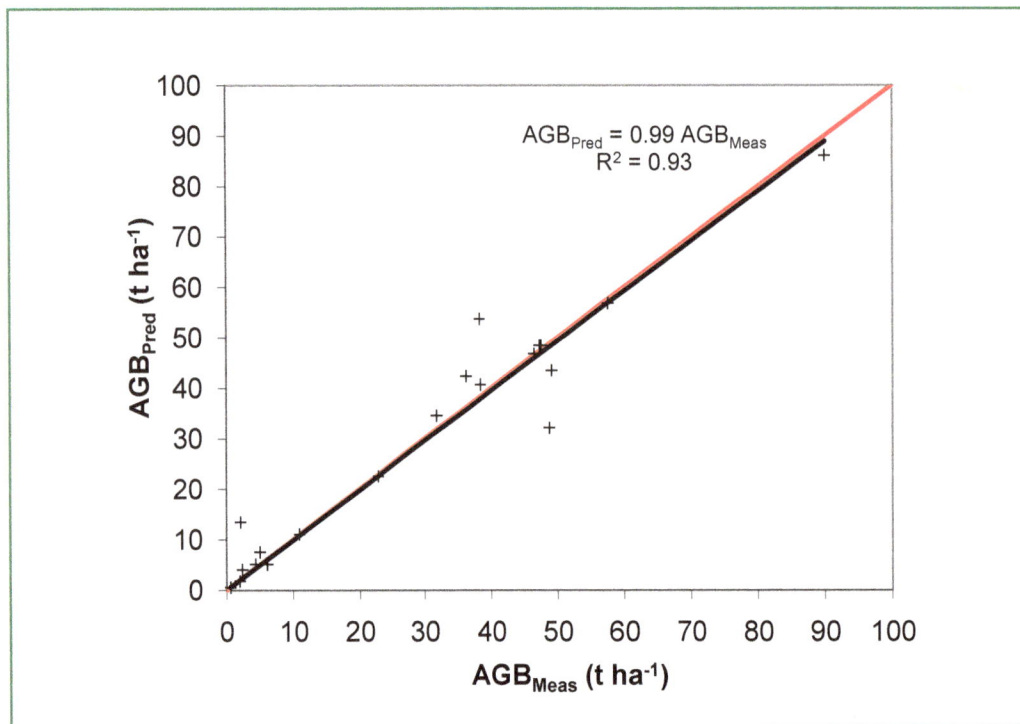

Figure 3.22 Comparison of site AGB calculated from field data (AGB_{Meas}) with values predicted using the Equations 3.8–3.11 (AGB_{Pred}). The lines show the line of best fit using a linear regression analysis (black) and the 1:1 relationship (red).

We noted above that the range of F_E for eucalypt woodlands in our field survey was from 0.22 to 0.32. The range of F_E for areas mapped as woodland in Figure 3.13, however, varies from 0 (no evergreen vegetation) to 0.5 (Figure 3.23). Those areas having very low values of F_E are under leasehold land tenure. As noted above (Section 3.2.5), most of the woodland under leasehold tenure has been subjected to mineral exploration and/or timber cutting—disturbances accounted for by scenarios (ii) and (iii). Low values of F_E can arise if regeneration of woodland has been suppressed by management actions that might also have resulted in further land clearing. We utilised Google Earth and high-resolution aerial photographs to investigate the density of woodland tree cover on land under leasehold tenure. One such area is shown in Plate 6. In this example, the woodland has evidently been thinned or cleared (in the south-western corner of the photo) for pastoralism. Plate 7 shows a further example of woodland on land that was formerly under pastoral leasehold tenure but is currently within a nature conservation reserve. The woodland has been thinned rather than cleared. The range in F_E for woodlands over the GWW (max. 0.5) exceeds the F_E values for the field sites (max. 0.32); hence predictions of AGB will exceed the calibration data for the equations. For other woody vegetation types, the range in F_E at the field sites is similar to the full range over the GWW.

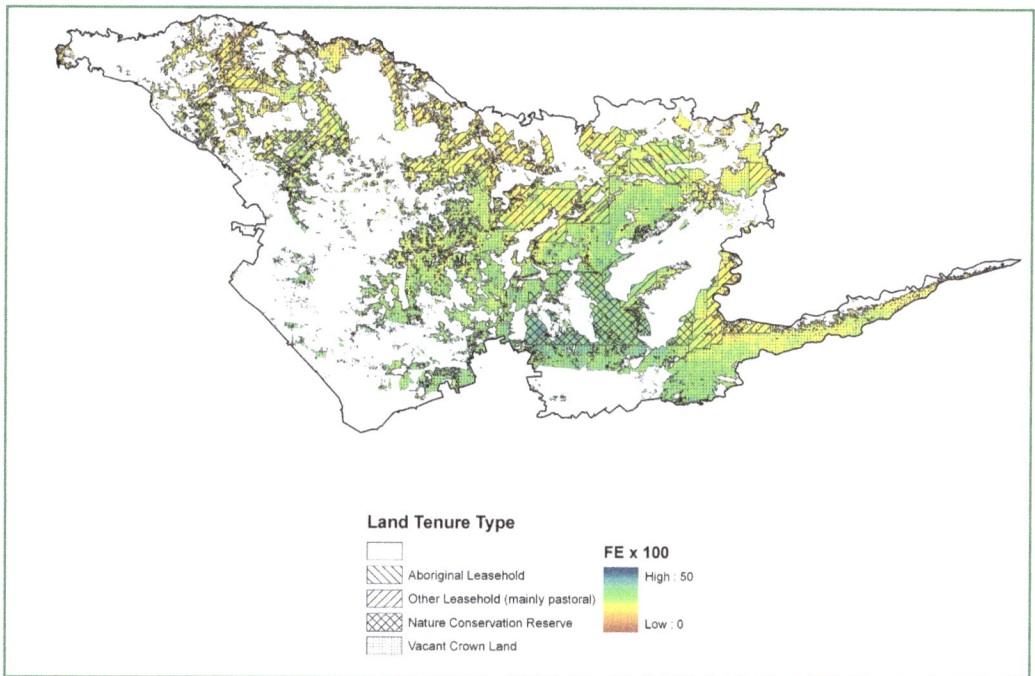

Figure 3.23 F_E of extant eucalypt woodland in 2008, overlayed with land tenure type (see Table A1—Land Tenure). Note the lower values of F_E for woodland under leasehold land tenure.

This figure incorporates Australian land tenure (1993) data that are Copyright Commonwealth of Australia 1993.

Plate 6 Aerial photograph showing a region mapped as woodland under leasehold land tenure in the GWW, north-east of Kalgoorlie. The region has low F_E (F_E = 0 in some MODIS pixels) and expanses of bare ground are clearly visible. A dam is visible in the south-western quarter.

Reproduced by permission of WA Land Information Authority, CL34/2010 (<www.landgate.wa.gov.au>).

Plate 7 Salmon Gum (*Eucalyptus salmonofloia*) woodland on Jaurdi Station, a former pastoral lease now managed by WA Department of Environment and Conservation.

4. SPATIAL BIOMASS AND CARBON ESTIMATES FOR THE GWW, 2008

Based on the methods and data detailed above, estimates of biomass and carbon content for the GWW are presented in Figure 4.1 and Table 4.1. We have assumed that carbon makes up 50 per cent of plant dry-mass or biomass (Gifford 2000). The four equations in the previous section were used for the categories of: undisturbed woodland (Figure 3.13); woodland disturbed by timber cutting (regardless of mineral exploration; Figure 3.15); woodland not disturbed by timber cutting but disturbed by mineral exploration (all uncut woodland on granite-greenstone lithology); and other woody vegetation of heath, shrub, marlock, mallee and low open woodland (Figure 3.13). In cases where both mineral exploration and timber cutting have occurred, the equation for timber cutting (Equation 3.10) was applied, as this represents the greater disturbance. AGB was calculated from the F_E value for every pixel within the area for each category. BGB and AGB_{dead} were then calculated from AGB. The sum of biomass carbon for the area of each category gave the total biomass carbon for the area mapped as the GWW to be 312 Mt. This value represents the current carbon stock inclusive of the impacts of human activities and natural disturbance regimes. Table 4.1 gives the estimated current biomass and biomass carbon stock for the GWW, while in Table 4.2 the values for the four vegetation categories are shown. A map showing the distribution of AGB is presented in Figure 4.1. In this figure, we also provide histograms showing the frequency distribution of grid cells within size classes of AGB predicted by our models for the present (2008) vegetation and the 'no-disturbance scenario' (Chapter 5). We have not provided maps of the other biomass components as these have a linear relationship to AGB.

We compared the GWW total biomass carbon stock estimate with an estimate derived from the continental analysis of Berry and Roderick (2006) of 437 Mt. The latter estimate was derived using global empirical relationships between fluxes (net primary productivity and gross primary productivity) and carbon stocks in stems, roots and leaves. The GWW values were extracted from the spatial layer of continental carbon stocks. Berry and Roderick assumed that the GWW vegetation was in its climax successional state and did not account for the removal of biomass carbon through timber cutting. The inputs included a data set of F_E calculated from the 1981–91 monthly time-series of NDVI at ~5 km × 5 km spatial resolution sourced from NOAA AVHRR[10] satellite imagery. F_E is an input variable into the Berry and Roderick model and our spatial estimates, but the F_E spatial layers for the two models were derived from different satellite data sets—different satellite sensors and different time periods, 1981–91 for the Berry and Roderick model. We tested the possibility that differences in F_E could be a major driver of differences in model outputs. We re-projected the F_E spatial data layer of Berry and Roderick (2006) to match the projection and grid-cell size of our MODIS imagery, then created a difference image by subtracting the mean F_E for the 12-month period (from July 2006 to June 2007) from the re-projected Berry and Roderick image. This difference image is shown in Figure 4.2. This revealed that over large areas of the GWW there is little difference in F_E between the 1981–91 period and our 2006–07 imagery. In some parts of the GWW, however, F_E clearly is lower at present than it was two decades ago. To assess the reasons for this change, we show the fire footprints and land tenure as overlays in Figure 4.2. Major reductions in F_E appear to result from fire, but land clearing associated with livestock grazing and mining also appears to have reduced F_E in the northern GWW where fire is absent.

---

10   National Oceanic and Atmospheric Administration Advanced Very High Resolution Radiometer.

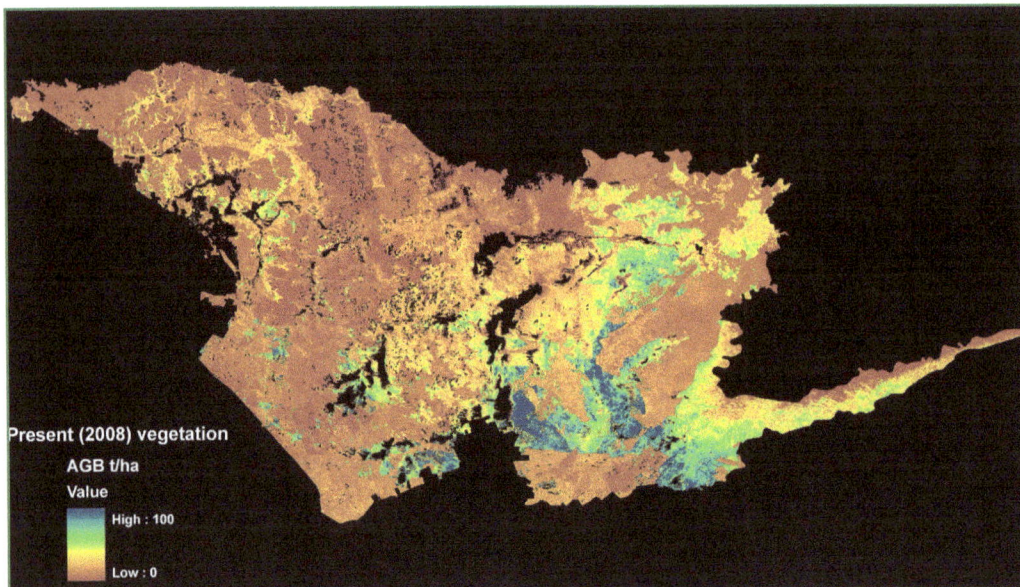

Figure 4.1a

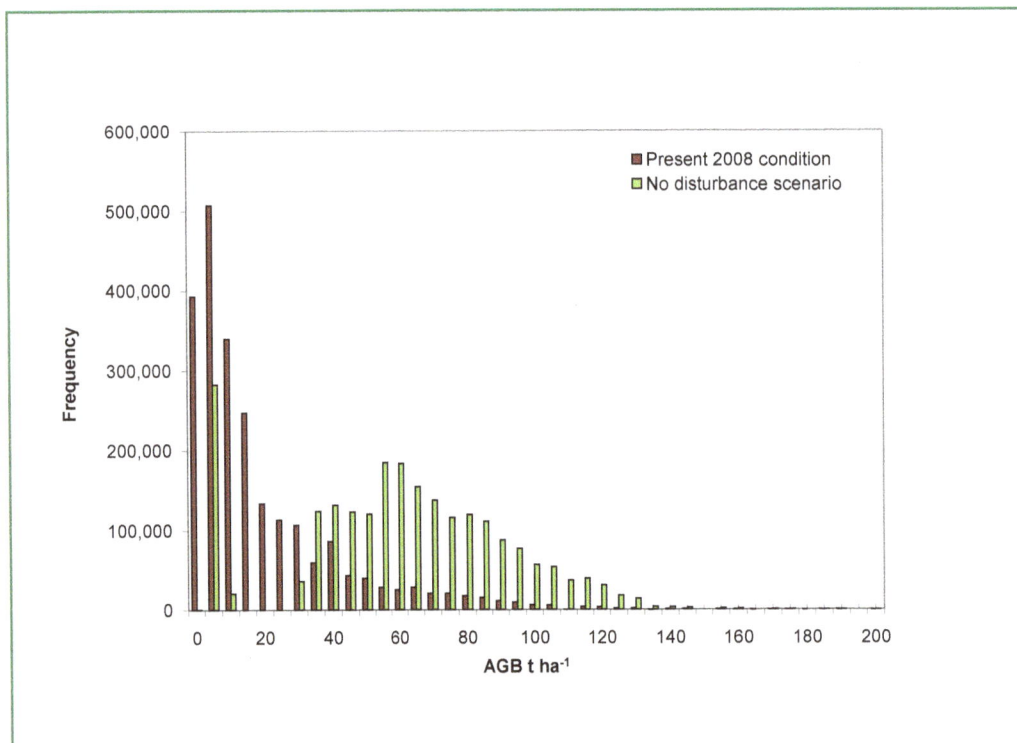

Figure 4.1b

Figures 4.1 (a) Map showing estimated above-ground biomass (AGB, t ha⁻¹) for the GWW based on vegetation structure shown in Figure 3.13. (b) Histogram showing the frequency distribution of 250 m × 250 m grid cells within size classes of AGB for the present (2008) vegetation (see Figure 3.13) and the predicted AGB for the 'no-disturbance scenario' (see Figure 3.14). The AGB carbon is equivalent to 50 per cent of the AGB and it has units of t C ha⁻¹.

Figure 4.2 Difference image showing the change in FE between 1981–91 (Berry and Roderick 2006) and 2006–07. Red areas indicate a reduction in greenness since 1981–91 and green areas indicate an increase. Cream areas show no change. Single hatches indicate fire footprints mapped over the period 1991–2007. Cross-hatching identifies the location of land under leasehold and freehold land tenure types.

The estimated mean AGB carbon of long-undisturbed woodland (category [i] in Table 4.2) is 24 t ha⁻¹ while the mean living biomass carbon (AGB + BGB) is 38 t ha⁻¹. These values fall within the range of biomass carbon estimates reported for studies in other Australian woodlands (Table 4.3). We find, however, that the carbon density (t C ha⁻¹) of living biomass in the long-undisturbed woodlands of the GWW is just 13 per cent of the estimated mean carbon density in living biomass of long-undisturbed forests of south-eastern Australia (Table 4.3). This is to be expected, as the south-eastern forests occur where there is much greater water availability throughout the year (Specht 1981) and consequently there are more trees per hectare having larger stem diameters (dbh) and taller stems. All else being equal, we would predict from Equations 3.8 and 3.9 that AGB of woodland affected by mineral exploration (that is, on greenstone), but not affected by timber cutting (category [ii] in Table 4.2), would be ~70 per cent of the value for the long-undisturbed woodland (category [i]), and this is the case. Similarly, Equations 3.8 and 3.10 predict that the AGB of woodland affected by timber cutting (category [iii] in Table 4.2) should be ~52 per cent of the category (i) value. We find, however, that the category (iii) value is just 46 per cent of the category (i) value. A substantial area that has been subjected to timber cutting previously is now leasehold land tenure and, as we noted above, the woodland appears to have been partly cleared or thinned. Consequently, the carbon stock has been reduced more than would be expected from timber cutting alone.

For shrublands and non-eucalypt dominated woodlands (category [iv]) that now cover ~54 per cent of the GWW, the estimated mean AGB carbon is 4 t ha⁻¹. This is comparable with estimates obtained for heath by several studies in southern Australia and for eucalypt re-growth in semi-arid Queensland (see Table 4.4). Our estimate of 3 t C ha⁻¹ for BGB is, however, much lower than published estimates of BGB of heath (Table 4.4). Thus, we could have underestimated the BGB of category (iv) vegetation, though this would need to be confirmed by field studies.

These results for the spatial estimation of biomass for the GWW are limited by the available data and information used as input to the models. The 21 field sites provided limited calibration data given the large variation in vegetation type and disturbance classes. It is also important to be clear on the assumptions used in our modelling.

- We assumed that the trees in the GWW are similar in wood density, structure, degree of piping by termites, and so on, to the trees from which the applied allometric equation was derived.

- We identified two major types of disturbance in woodlands—timber cutting and mineral exploration—and we applied individual equations that related canopy greenness to AGB (Equations 3.9 and 3.10) to account for the impact of each of these types of disturbance on the site AGB. We do not know precisely the extent to which the biomass was impacted, or has recovered, within the area mapped as being disturbed. We do not specifically model the impacts of pastoralism. Where woodland under leasehold land tenure has been disturbed by mineral exploration and/or timber cutting, we applied Equations 3.9 and 3.10. We assumed that any additional woodland thinning or clearing due to pastoralism was adequately accounted for in these equations through a reduction in canopy greenness (F_E).

- We do not have reliable allometric equations for non-tree woody species. Rather, we used an estimated geometric function for which field validation is necessary. We note, however, that in studies of forest vegetation biomass carbon, the shrub component is commonly not estimated as it is very small in comparison with the tree biomass.

- Boundaries of the disturbance categories of timber cutting and mineral exploration were derived from available maps. The timber-cutting map shows the general area of cutting but no information about intensity or volumes of timber extracted. The mineral exploration impacts were based on the 1:250 000 scale surface lithology layer (see Figure 2.6). We assumed that all areas within the mapped granite-greenstone belt have been equally impacted by mineral exploration.

- We modified the NVIS MVG vegetation map layer to take into account the impacts of fire on AGB. Areas mapped as eucalypt woodland (MVG 5) on the NVIS extant vegetation map, which have since been burnt in intense fires, were reclassified as low closed forests and tall closed shrublands (MVG 15) on our map of the vegetation structure of the GWW in 2008. We have assumed that intense fire kills all large eucalypt trees in the GWW and that they are replaced with a dense thicket of regeneration arising from seeds or root suckers. This assumption is consistent with published accounts and our analyses but needs to be further validated by additional field investigations and analyses.

- BGB and AGB_{dead} are calculated as proportions of AGB. Thus, any errors in the estimation of AGB will be carried through into estimates of these other biomass components. We have no site measurement data to validate estimates of these components.

Table 4.1 Estimated biomass and biomass carbon for the GWW. The area of vegetation cover over which the calculations are based is 14 926 197 ha. Three sets of values are given: present (2008) condition is the 'best estimate' based on current knowledge; no-disturbance scenario is an estimate of 'what might have been' if the woodlands had not been impacted on by fire, timber cutting, mineral exploration and pastoral land management; the Berry and Roderick 'pveg' is derived from their estimates (Berry and Roderick 2006) of biomass carbon for the Australian continent based on NOAA AVHRR satellite data from 1981 to 1991 and global empirical equations.

Present 2008 condition		Average biomass t ha^{-1}	GWW total Mt	Average biomass carbon t ha^{-1}	GWW total carbon Mt
AGB	From Equations 3.8–3.11	21.4	320	10.7	160
BGB	AGB × 0.6	12.8	192	6.4	96
Live biomass	AGB + BGB	34.3	511	17.1	256
AGB$_{dead}$	AGB × 0.35	7.5	112	3.8	56
AGB$_{total}$	AGB + AGB$_{dead}$	28.9	432	14.5	216
Total biomass	AGBtotal + BGB	41.8	624	20.9	312
Berry–Roderick pveg					
AGB	Live biomass × 0.625	30.0	448	15.0	224
BGB	Live biomass – AGB	18.0	269	9.0	134
Live biomass	From Berry and Roderick	48.0	716	24.0	358
AGB$_{dead}$	AGB × 0.35	10.5	157	5.3	79
AGB$_{total}$	AGB + AGB$_{dead}$	40.5	605	20.3	303
Total biomass	AGBtotal + BGB	58.5	874	29.3	437
No-disturbance scenario (that is, what might have been)					
AGB	From Equations 3.8 and 3.11; F$_E$ from W (Equations 5.1 & 5.2)	62.9	938	31.4	469
BGB	AGB × 0.6	37.7	563	18.9	281
Live biomass	AGB + BGB	100.6	1 501	50.3	751
AGB$_{dead}$	AGB × 0.35	22.1	330	11.0	165
AGB$_{total}$	AGB + AGB$_{dead}$	84.9	1 268	42.5	634
Total biomass	AGB_total + BGB	122.7	1 831	61.3	915

Table 4.2 Estimated biomass carbon in 2008, for each of the modelled vegetation categories.

	Modelled class							
	(i) No timber cutting, no mineral exploration disturbance		((ii) Mineral exploration, no timber cutting		(iii) Timber cutting		iv) Shrublands and non-eucalypt dominated woodlands	
Area (× 1000 ha)	3327		2319		1269		8012	
Carbon component	t C ha⁻¹	Mt C	t C ha⁻¹	Mt C	t C ha⁻¹	Mt C	t C ha⁻¹	Mt C
AGB mean [standard deviation]	24 [14]	79	17 [11]	22	11 [6]	25	4 [3]	34
BGB mean	14	47	10	13	6	15	3	20
Living biomass	38	127	27	40	17	35	7	54
Total biomass (living + dead)	46	155	33	48	20	42	8	54
Soil	38	125	39	91	39	49	42	339
Total carbon stock	**84**	**280**	**72**	**139**	**60**	**91**	**51**	**406**

Table 4.3 Comparison of carbon stocks for the GWW and estimates published by other studies for Australia woodlands[11].

Vegetation type and location	Author	Carbon stock (t C ha⁻¹) mean [standard deviation]		
		AGB	**BGB**	**Living**
Temperate eucalypt woodland, no timber cutting, no mineral exploration disturbance GWW	This study	24 [14]	14	38
Tropical savanna woodland	Cook et al. (2005) Kapalga, NT			22.9 [7.8] shallow soil 43.5 [6.6] deep soil
Tropical savanna woodland	Chen et al. (2003)	30.7 [7.3]	19.3 [12.6]	50 [19.9]
Black box woodland, Victoria	Grierson et al. (1992)	25.0		
Eucalyptus crebra intact woodland, central Queensland	Burrows et al. (2000)	56.7 [13]		
Eucalyptus melanophloia intact woodland, central Queensland	Burrows et al. (2000)	20.7 [4.7]		
Eucalyptus populnea intact woodland, central Queensland	Burrows et al. (2000)	35.2 [4.9]		
Eucalypt on hills, mature; semi-arid Queensland (rainfall 460–780 mm yr⁻¹). Mean canopy cover 32.4%	Fensham et al. (2002)	36.0		
Eucalypt on clay, mature; semi-arid Queensland (rainfall 460–780 mm yr⁻¹). Mean canopy cover 16.9%	Fensham et al. (2002)	18.3		
Eucalypt on texture-contrast soils, mature; semi-arid Queensland (rainfall 460–780 mm yr⁻¹). Mean canopy cover 24.9%	Fensham et al. (2002)	34		
Eucalypt on sand, mature; semi-arid Queensland (rainfall 460–780 mm yr⁻¹). Mean canopy cover 29.3%	Fensham et al. (2002)	29.7		
Eucalyptus populnea open woodland, central-west NSW	Harrington (1979)	27.3		
Carbon carrying capacity, south-eastern Australia forests	Mackey et al. (2008)			289 [226]

---

11   See Raison et al. (2003) for a summary of biomass estimates.

Table 4.4 Comparison of carbon stocks for GWW shrubland vegetation and estimates published by other studies for Australian shrublands[12].

Vegetation type	Author and location	Carbon stock (t C ha⁻¹) mean [standard deviation]		
		AGB	BGB	Living
Temperate shrubland, GWW	This study	4 [3]	3	7
Eucalypt on texture-contrast soils, re-growth <5 m tall; semi-arid Queensland (rainfall 460–780 mm yr⁻¹). Mean canopy cover 5.8%	Fensham et al. (2002)	1.7		
Eucalypt on sand, re-growth <5 m tall; semi-arid Queensland (rainfall 460–780 mm yr⁻¹). Mean canopy cover 14%	Fensham et al. (2002)	2.1		
Acacia on clay, re-growth <5 m tall; semi-arid Queensland (rainfall 460–780 mm yr⁻¹). Mean canopy cover 9.1%	Fensham et al. (2002)	2.2		
Heath, south-eastern South Australia	Specht et al. (1958)	13.2		
Heath, south-eastern South Australia, 12 years post-fire	Specht (1966)	4.6		
Heath, south-western WA	Low and Lamont (1990)	7.0	15.4	22.7
Heath, Frankston, Victoria	Jones (1968)	4.8 5.1	30 37	34.8 42.1
Heath, Wilsons Promontory, Victoria	Groves (1965)	3.9	21	24.9
Heath, Wilsons Promontory, Victoria	Groves and Specht (1965)	6.8 9.0 14.0	6.6 17.3	13.4 26.3

---

12   See Raison et al. (2003) for a summary of biomass estimates.

5. BIOMASS AND CARBON ESTIMATES: HYPOTHETICAL NO-DISTURBANCE SCENARIO

Our analyses reveal that the vegetation over significant areas of the GWW has been highly disturbed by direct and indirect human impacts (Figure 3.18) and consequently is below its carbon carrying capacity. The disturbance factors of changed fire regimes (primarily due to increases in human ignition), timber cutting and mining appear to have altered the vegetation structures such that the extant vegetation cannot be correlated with, and predicted from, environmental factors alone, as would be expected for natural ecosystems. As we noted in Sections 2.4 and 3.22, the published maps of pre-disturbance condition (AUSLIG 1990 and NVIS 3.0, DEWHA 2005) do not adequately take into account the disturbance to the vegetation of the GWW that occurred before the initial vegetation mapping. We determined that the best means of estimating vegetation structure, and hence its carbon carrying capacity, in the absence of disturbance by human activities was to assume a vegetation cover of woodland in all areas with potentially suitable environmental conditions, as is shown in Figure 3.14. To estimate the AGB for this scenario, we applied the equations derived for undisturbed vegetation (Equations 3.8 and 3.11).

Fire and land use have, however, clearly impacted on F_E (see Figure 4.2). Thus, before applying these equations, we needed a better estimate of F_E for undisturbed vegetation. In a study of the relationship between F_{PAR} and climate over the Australian continent, Berry and Roderick (2002) found that there was a linear relationship between F_{PAR} and W—an index of water availability—along the aridity gradient (Equation 5.1).

Equation 5.1

$W = P - R_S/L$

In Equation 5.1, P (mm yr^{-1}) is the average precipitation, R_s is the sum of the global solar irradiance over the annual period (MJ m^{-2} yr^{-1}) and L is the amount of energy (joules) required to evaporate 1 kg of liquid water. (Note that 1 kg of water is equivalent to a 1 mm thick layer of water over a square metre of surface.) A map showing the distribution of W over the GWW is given in Figure 5.1. As W has negative values, we added a constant—3000—to create the variable W_{3000} for subsequent analyses.

In order to relate W_{3000} to F_E, we first identified those grid cells within the GIS layer of F_E that represent long-undisturbed woodland vegetation (F_{Eluv}). We then plotted F_{Eluv} against W_{3000} and derived the equation of best fit (Equation 5.2).

Equation 5.2

$F_{Eluv} = 0.03W_{3000} + 18$

Following that, we created a spatial layer that comprised the maximum value of F_{Eluv} predicted from Equation 5.2 and actual F_{Eluv} (Figure 5.2). Finally, we calculated AGB for undisturbed vegetation using Equations 3.8 and 3.11. A summary of the biomass estimates for the GWW under the no human disturbance scenario is included in Table 4.1 and a histogram showing the frequency distribution of size classes of AGB 'no-disturbance scenario' is shown in Figure 4.1. Biomass carbon for this scenario was 915 Mt for the GWW—approximately three times the estimated present biomass. In Figure 4.1, the low values of AGB (that is, less than 15 t ha^{-1}) are associated with non-eucalypt dominated vegetation groups (MVGs 6, 7, 8, 17, 18, 20, 22). We have no field measurements of biomass of undisturbed MVGs 6, 7 and 8 to test the veracity of these estimates. We speculate, however, that our modelled estimates understate the AGB of undisturbed *Acacia* (mulga, MVG 6), *Callitris* (MVG 7) and *Casuarina* (MVG 8) woodlands in the GWW.

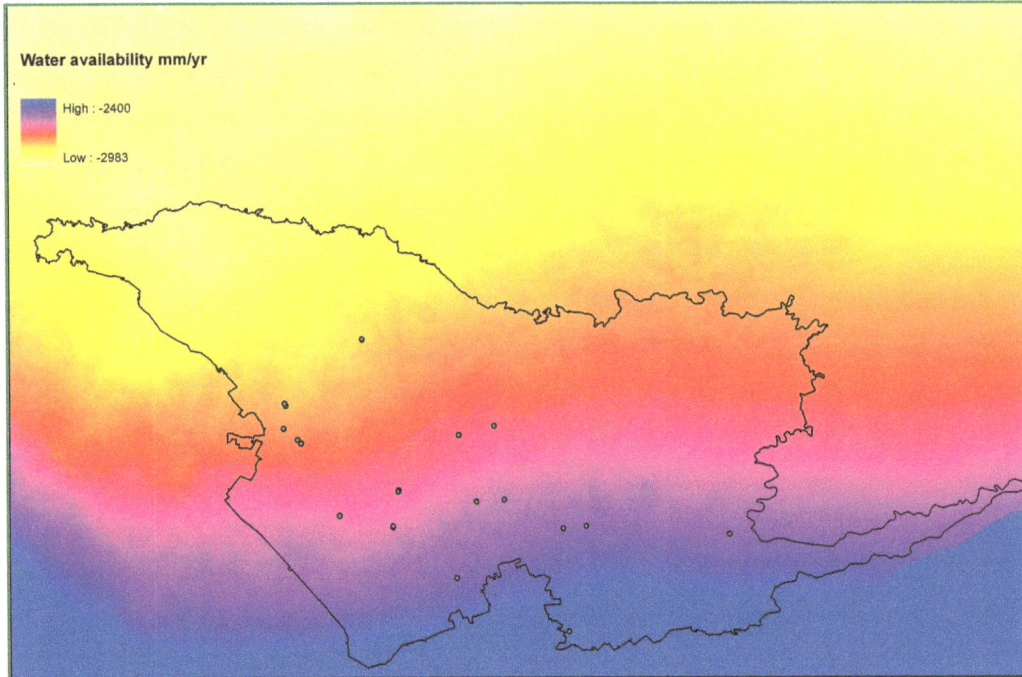

Figure 5.1 Map showing spatial variability of a climatic index of water availability (W) over the GWW. The locations of the ANU field sites are indicated by the green circles.

For source of spatial data inputs for calculation of W, see Table A1—R_s and P.

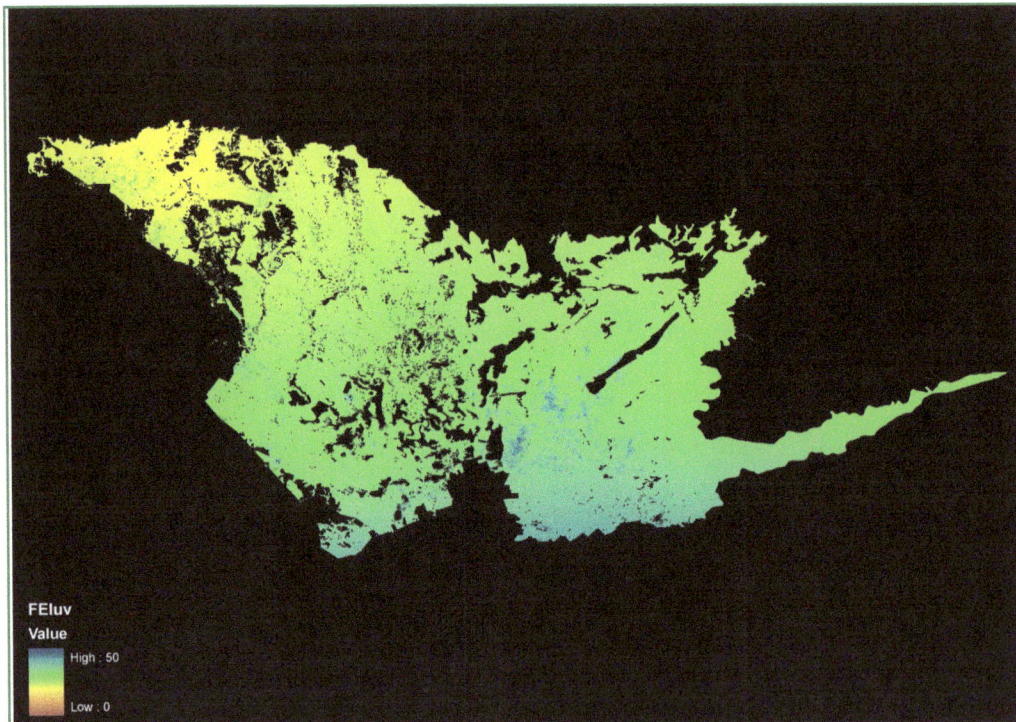

Figure 5.2 Map showing estimated F_{Eluv}, F_E of vegetation if it were in the long-undisturbed condition, estimated from the water availability index, W (see Equation 5.1).

6. SOIL CARBON

Soil is widely recognised as containing the largest pool of terrestrial carbon (Jobbágy and Jackson 2000) and components of the soil organic carbon pool (SOC) have great longevity—up to centuries or millennia. There are, however, very few studies of SOC in forest and woodland ecosystems in Australia. A study by Skjemstad et al. (1996), which covered a range of soil types, revealed that charcoal made up to 30 per cent of the SOC in the Australian soils they sampled. Other constituents having great longevity included humic acids and lignins.

There were no data for SOC for the GWW, so we needed to use a modelling approach to provide an estimate and complete the carbon budget. We identified two publicly available spatial data sets: the Australian Soil Resource Information System (ASRIS) (CSIRO 2007) and SOC data from the field study by Wynn et al. (2006).

The ASRIS spatial data set provides only partial coverage of SOC in the GWW, but it does have full coverage of estimated soil thickness and bulk density of the A and B horizons. Where coverage exists, SOC of the A and B soil horizons was calculated (Figure 6.1).

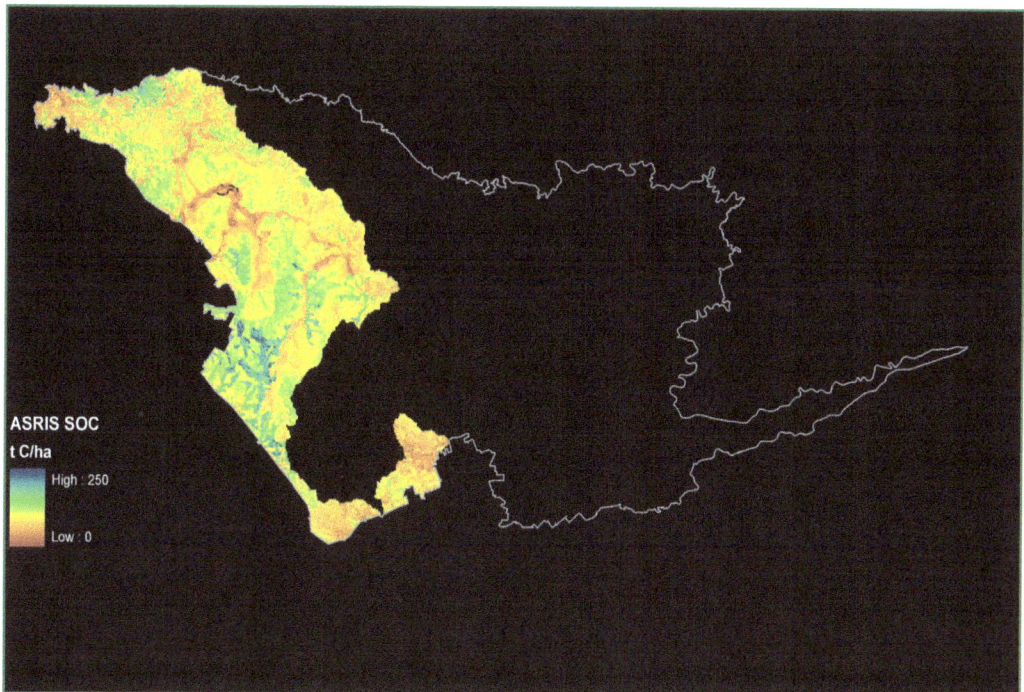

ASRIS SOC
t C/ha
High : 250
Low : 0

Figure 6.1 Total soil organic carbon derived from ASRIS (CSIRO 2007) spatial data layers.

Data source: SOC—Table A1.

The field survey of Wynn et al. extended over a wide range of environments within 48 regions of Australia, but not in the GWW. They used 7050 soil cores to determine SOC for the top 30 cm of soils beneath trees and grasses for the 48 regions. They described the relationship between SOC (sandy soils) and the water availability index, W (see Equation 5.1), as a sigmoidal function. A supplementary table to Wynn et al. (2006) includes organic carbon fraction (fracC) values for the 48 regions. Using an average of the values of fracC in the top 30 cm obtained from sites with trees and grass, and excluding rainforest, pasture and graminoid (buttongrass-dominated) vegetation types, we derived the power relationship (Equation 6.1).

Equation 6.1

$$fracC_{Wynn} = 0.00000002W_{4000}{}^{1.736}; R^2 = 0.73, n = 39$$

in which

$$W_{4000} = W + 4000 \text{ mm yr}^{-1}$$

The index of water availability, W, is defined in Equation 5.1. (Note: in Section 5, we calculated W_{3000} as the addition of 3000 was sufficient to create a variable with positive values for the GWW. As Wynn et al. use W_{4000} for their continental data set, we use this value to apply to analyses with their data.) We applied Equation 6.1 to our spatial data layer of W_{4000} to obtain spatial estimates of fracC in the A horizon. We compared these spatial estimates ($fracC_{Wynn\_A}$) with the values in the ASRIS data layer ($fracC_{ASRIS}$) using a linear regression analysis. The line of best fit was described by Equation 6.2.

Equation 6.2

$$fracC_{Wynn\_A} = 0.000023 + 0.366243 \, fracC_{ASRIS\_A} \, ; R^2 = 0.90$$

Although the spatial layers are correlated, the estimates based on the data of Wynn et al. are on average 37 per cent of the values given in the ASRIS database. The derivation of the ASRIS estimate and its source data are not known, thus we cannot conclude whether the ASRIS or the Wynn approach is preferable, except that ASRIS does not give coverage of SOC for the whole GWW.

We used the spatial estimates of $fracC_{Wynn\_A}$ along with the ASRIS horizon thickness and bulk density to estimate SOC of the A horizon. Because the organic carbon is derived mostly from near-surface and above-ground plant parts, the fraction of organic carbon in the A horizon is greater than that in the B horizon. Consequently, we needed to adjust $fracC_{Wynn\_A,}$ in order to estimate SOC in the B horizon. We find that for the area covered by ASRIS, the mean ratio of $fracC_{ASRIS\_A}$: $fracC_{ASRIS\_B}$ is 2.13. Applying this ratio, we have Equation 6.3.

Equation 6.3

$$fracC_{Wynn\_B} = \frac{fracC_{Wynn\_A}}{2.13}$$

We used this spatial estimate of the fraction of SOC in the B horizon along with the ASRIS spatial estimates of the B horizon thickness and bulk density to calculate SOC for this soil layer. The estimates of SOC in the A and B horizons are presented in Figure 6.2 and Table 6.1.

The resulting estimated SOC—637 Mt C for the GWW or 40 t C ha^{-1}—was constant across all four vegetation categories (Table 4.2). This is approximately twice as much carbon as we estimated to make up the total biomass in 2008 (see Table 4.1). We noted above that the carbon content estimated from the measurement data of Wynn et al. (2006) is about 37 per cent of the estimate based on the ASRIS layers. Thus, we might expect that, if ASRIS coverage for the entire GWW was available, it would result in an estimate of 1722 Mt C. This is almost twice as much carbon as we estimated to make up the total biomass in the hypothetical 'no-disturbance' scenario.

Figure 6.2 Map showing the estimated distribution of soil organic carbon (SOC, t C ha^{-1}). Calculations were based on the relationship between the fraction of organic carbon in the soil and the water availability index, W, using the data provided as supplementary material to Wynn et al. (2006). The depths and bulk densities of the uppermost soil horizons (A and B) are from ASRIS (CSIRO 2007) spatial data layers.

Data sources: Table A1.

Harms et al. (2005) found that soil carbon stocks at uncleared sites in the semi-arid rangelands of central and southern Queensland averaged 41 per cent of the mean total C (soil C plus biomass C). This corresponded to an average soil carbon stock for the top metre of soil of 62.5 t C ha^{-1}. In contrast, the SOC$_{Wynn}$ estimate averages 67 per cent of the mean total C stock in the GWW. As the live biomass in the GWW has been greatly reduced over recent decades due to tree kill by fire, a greater proportion of SOC to total C might be expected.

Table 6.1 Estimates of mass of soil organic carbon in the GWW. The estimated ASRIS A+B horizons value is based on the relationship given in Equation 6.2, which shows that, for the part of the GWW covered by ASRIS, our estimate based on the data of Wynn et al. is 37 per cent of the ASRIS value.

Method—soil horizon	Average t C ha^{-1}	Total for GWW Mt C
Wynn A –horizon	20.4	325
Wynn B –horizon	19.7	314
Wynn A+B horizons	40.1	639
ASRIS A+B horizons = (Wynn A+B horizons) × 2.7	108.0	1 725

7. MANAGEMENT OPTIONS TO PROTECT AND RESTORE THE GWW'S CARBON STOCKS

In this report, we define eucalypt woodlands as vegetation dominated by eucalypt trees exceeding 10 m in height and with a projective foliage cover[13] of 10–30 per cent. The definition of eucalypt forests differs from woodlands only in that the projective foliage cover is >30 per cent (Table 2.1). At the time the GWW was first explored by European people, in expeditions led by Roe in 1849 and Lefroy in 1863 (Beard 1968), the region probably supported a far more extensive cover of eucalypt woodland and forest than it does now. If these woodlands were in an undisturbed condition, and had remained so to the present day, it is possible that the woodland structural formation would have a distribution similar to that depicted in Figure 3.14, with the woodland extending over 12.9 million ha, or 80 per cent of the GWW, with biomass carbon of ~900 Mt. This 'no-disturbance scenario' model-derived estimate is based on an assumption that none of the woodland is in a pyric successional state (that is, recovering from fire), and it should be treated as an estimate of the maximum possible store of biomass carbon in the GWW if intense fire had been excluded for centuries. The distribution of woodland in Figure 3.14 is similar to, but more extensive than, that mapped in the Natural Vegetation Map (Figure 2.9a) and the NVIS map (Figure 2.10) because our analyses indicate that these map layers show pyric successional vegetation structural formations arising from fires that have occurred in the period following the advent of explorers, prospectors and timber cutters in the GWW.

We estimate, however, that at present (2008) eucalypt woodland extends over just 6.9 million ha, or 43 per cent, of the GWW, and the total biomass carbon of the GWW is ~300 Mt. The 6 million ha that might once have had a vegetation cover of eucalypt woodland currently support a cover of eucalypt or acacia-dominated shrubland (NVIS MVGs 13–16). We attribute the change in vegetation structure primarily to changes in fire regimes: the advent and continuance of intense and more frequent fires in this region following the discovery of gold and subsequent settlement by European people. Timber cutting and the removal of vegetation for roads, seismic lines and excavations have also impacted on the biomass carbon stock. Our analyses, presented in Figure 3.8, indicated that during the period 1972–2007, 3.5 million ha of the GWW have been burnt by intense fire once, while 852 000 ha have been burnt twice and 23 000 ha have been burnt three times. In total, 4.4 million ha representing 30 per cent of the GWW have been burnt by intense fire over the past 36 years and now support pyric successional vegetation structural formations.

If fire were to be suppressed over the coming decades to centuries so that self-thinning could occur, the shrublands of seedling and re-sprouting eucalypts could re-grow into eucalypt woodland (MVG 5) with the proviso that the availability of water, nutrients and propagules post-fire is sufficient. From a management perspective, it can be argued that fire suppression will lead to fuel accumulation and subsequently to 'monster' fires (Bond and Keeley 2005). Thus, fire can be seen as a consumer of biomass that would otherwise accumulate indefinitely. This, however, disregards the role of wood-eating termites—voracious consumers of deadwood in the GWW. Several species of debris and wood-eating termites occur in the GWW (Hadlington 1987).

---

13   Projective foliage cover (PFC) is the proportion of the ground that would be shaded by the canopy on a clear day if the sun were directly overhead. Canopy cover is the proportion of the ground that lies beneath the crowns of the plants. PFC accounts for the gaps between the leaves in the canopy.

That termites prevent the excessive accumulation of biomass in the GWW is evident from the photos of sclerophyll woodland in Figure 2.8. In the following sections, we discuss the causes of fire in the GWW, the options for fire management and, finally, the likely time required for the post-fire succession to return to woodland.

7.1 FIRE IN THE GWW

Our results suggest that fire plays a major role in the diminution of biomass carbon in the GWW and any attempt to maximise biomass carbon stocks in the GWW necessarily requires a substantial reduction in fire frequency and intensity. Therefore, we need to consider the causes of the intense and extensive wildfires that now ravage the GWW.

Wildfire has three basic requisites: a supply of fuel, a source of ignition and an agent of spread. Landscape topography is also an important factor at many locations as slope angle affects the rate of spread and landform complexity can add to fire patchiness. As the GWW has relatively little topographic relief, we do not consider this factor any further.

Fuel, or combustible matter, is a spatially and temporally variable entity. Vegetation structure accounts for the spatial arrangement of fuel within a volume above the ground surface. The eucalypt woodlands of the GWW mostly have a sparse understorey and the trees are widely spaced (foliage projective cover of woodland is 10–30 per cent). For this reason, Beard (1968:252) noted that woodland 'is subject to burning to only a minor extent'. Yet, we have shown from Landsat and MODIS satellite data that 4.4 million ha of woodland in the GWW have been burnt in intense fires over the past 36 years, and suggest that a further 2 million ha could have been burnt earlier given the occurrence of pyric successional vegetation structures. In the seven years (2000–07) for which we have MODIS NDVI imagery (see Table A1—MODIS), intense fires burnt more than 2.6 million ha of the GWW. Therefore, we argue that extensive areas of eucalypt woodland (Table 2.1, class M2) have been replaced by eucalypt or acacia-dominated shrublands (Figure 3.9; Table 2.1, class S3)—vegetation structures that, according to Beard (1968), are subject to more frequent fire.

Temporal variation in fuel depends on the moisture and heat content of the living and dead biomass, and hence its potential to become fuel at a given time. Moisture and heat content of biomass are dependent largely on the ambient air temperature, relative humidity and rainfall (see Figure 7.1). Lower moisture content and higher temperature of the biomass reduce the amount of energy required to reach the ignition point. Across the GWW there is a marked seasonal cycle with the seasonality of rainfall greater on the western edge of the GWW (near Southern Cross and Hyden), a distinct winter maximum at these locations and less seasonality of rainfall elsewhere in the GWW. The combination of high daily temperatures and low rainfall during summer across the GWW results in low biomass moisture content and high biomass heat content and subsequently high availability of fuel.

The two basic sources of ignition of wildfires are natural and human. Lightning strikes undoubtedly ignite some fires within the GWW; however, over most of the GWW the number of thunder-days recorded in a year averages less than 20—in contrast with the Northern Territory tropical savanna woodlands, which commonly experience 60 or more thunder-days (Figure 7.2). In a continental-wide study of causes of vegetation fires in Australia, Bryant (2008) found that only 6 per cent resulted from natural ignition.

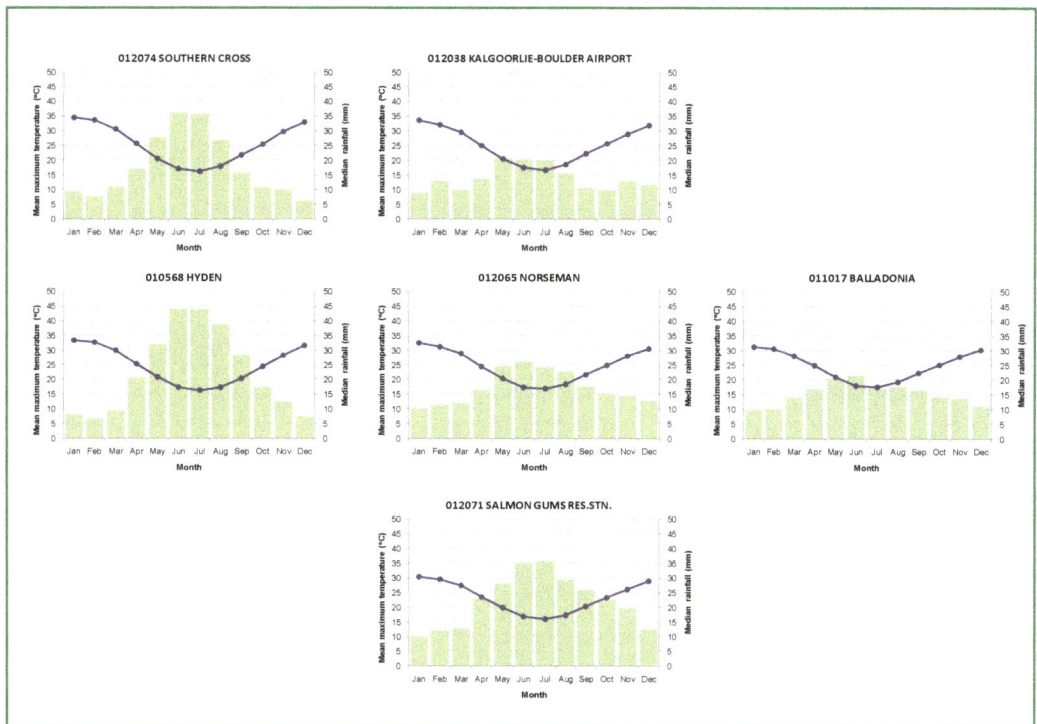

Figure 7.1 Mean maximum temperature (line) and median rainfall (columns) in the GWW region.

Data source: Bureau of Meteorology (<http://www.bom.gov.au/climate/data/index.shtml/>). This figure incorporates climate data that are Copyright Commonwealth of Australia 2009.

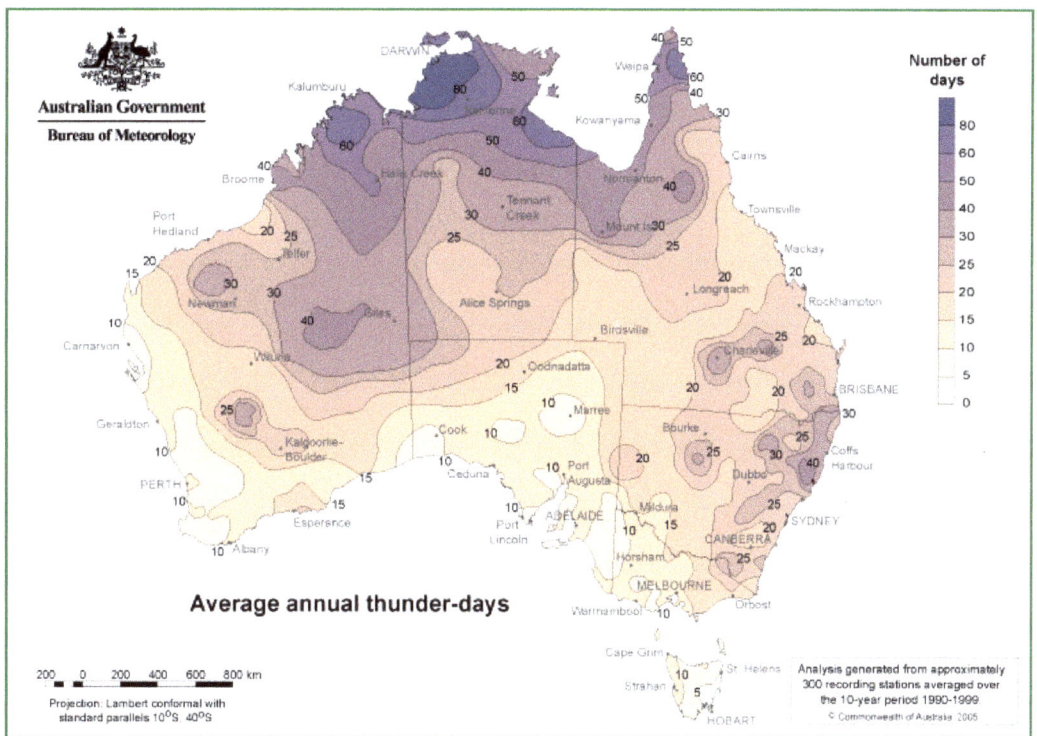

Figure 7.2 Average number of days (per annum) during which thunderstorms have been detected over a 10-year period (1990–99).

Data source: Bureau of Meteorology (<http://www.bom.gov.au/jsp/ncc/climate_averages/thunder-lightning/index.jsp>, updated 7 October 2009), Copyright Commonwealth of Australia, reproduced by permission.

Records of the Forests Department of Western Australia reveal that natural sources of ignition (lightning) historically caused from 1 to 6 per cent of fires in the south-western forests (Figure 7.3).

Fires ignited by humans can be accidental or deliberate. We define a deliberate fire as resulting from a person intentionally bringing burning material (such as a flaming match or an incendiary) into contact with fuel. Accidental ignitions can be further subdivided into: 1) passive fires ignited by, for example, fuel coming into contact with hot vehicle components, fuel heated by glass (discarded/broken bottles, and so on) that focuses solar radiation or flaming debris associated with blasting activities (NWCG 2005); 2) active fires ignited by careless use of burning materials—for example: cigarette butts or campfires. Deliberate fires can be subdivided into: i) legal ignitions—for example, prescribed burns and legal burn-offs; and ii) illegal ignitions—for example, illegal burn-offs and arson. All of these types of ignitions are reported to have caused fires in the south-western forests (Figure 7.3) and are likely to have also caused fires in the GWW.

Wind has an important role in spreading fire. It does this by bringing flames into contact with fuel, by providing a supply of oxygen, by removing moisture from the fuel and by transporting burning embers that can ignite spot fires up to 30 km distant from the fire front (Geoscience Australia 2009). Wind speeds of 12–15 km hr^{-1} and higher greatly impact on fire spread and intensity (Geoscience Australia 2009). Wind speeds above the 12–15 km hr^{-1} threshold are regularly attained in the GWW (Figure 7.4), although the one possible exception is the area closest to Hyden.

In summary, the fire conditions in the GWW consist of available fuel during hot, dry summers, wind as an agent of spread and some natural sources of ignition. The discontinuous fuel arrangement of original woodland vegetation might not generally have carried fire following natural ignitions before European settlement as the combination of conditions required for extensive wildfire events could have been lacking.

Because there is a lack of surface water in the GWW (see Table A2), Aboriginal people might have been absent from much of the area or present in only low densities during the hottest, driest, highest fire-danger months of the year. Historical observations of the habits of Aboriginal people of the GWW are scarce. In 1891, the Elder Exploring Expedition (Helms 1892–96) encountered only small groups of people as it moved through the western interior of Western Australia. Helms considered the people to be highly nomadic (and, in the Fraser Range region of the GWW, very undernourished) as a consequence of scant food resources. He observed that they were dependent solely on 'rock-holes' and 'native wells' for water during the dry season, which presumably limited their spatial range of movement. Rockholes are found mostly around granite outcrops. We assume that intentional burning by Aboriginal people would be limited to fires of low intensity and small spatial extent—for example, the firing of patches of spinifex grass (Triodia) to facilitate access to small marsupials that shelter within the tussock, as described by Helms (1892–96:255–6). The advent of mineral exploration, mining, the water pipeline, timber cutting, recreation and tourism has, however, resulted in a large increase in the potential sources of ignition that coincide with periods of high fuel availability and strong winds—conditions that allow intense, tree-killing fires to impact on the mature woodland vegetation. Subsequent post-fire successional changes in vegetation structure from woodland to dense shrubland (thickets of seedlings and root suckers) provide a more continuous fuel layer that is probably much more susceptible to fire than the woodland vegetation it replaces.

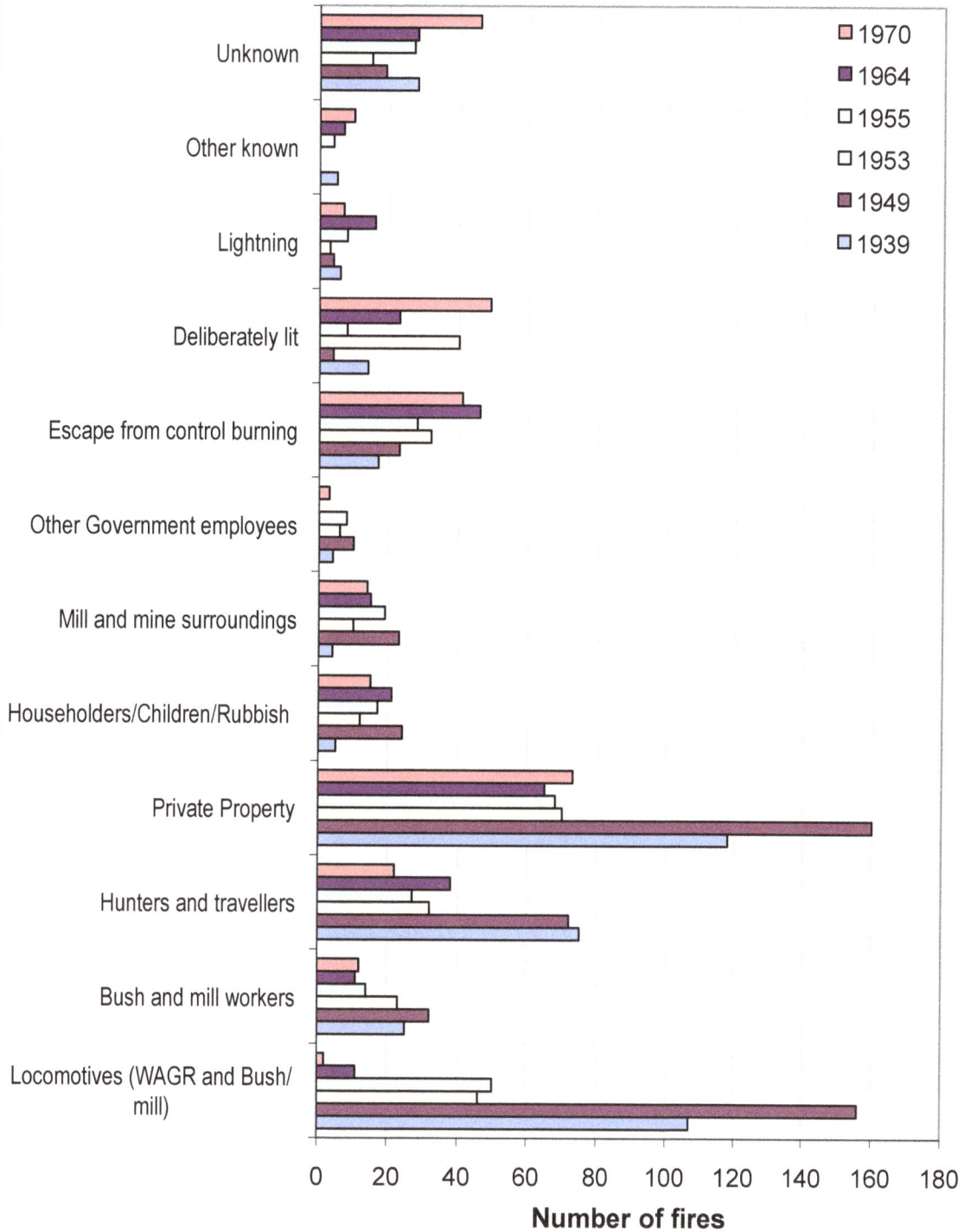

Cause of fires in south-western forests attended by Forest Department Western Australia

Figure 7.3 Historical causes of fire in the south-western forests of Western Australia.

Data source: Western Australia Forests Department (1934–70).

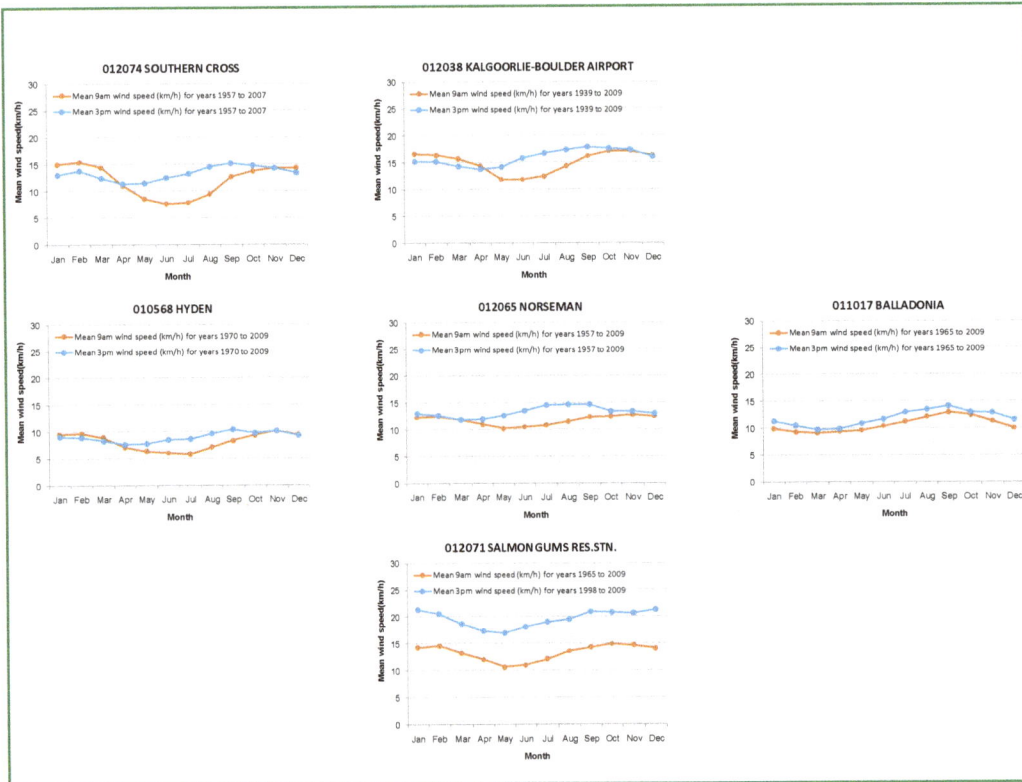

Figure 7.4 Mean wind speeds recorded in or near the GWW region.

Data source: Bureau of Meteorology (<http://www.bom.gov.au/climate/data/index.shtml/>). This figure incorporates climate data that are Copyright Commonwealth of Australia 2009.

7.2 OPTIONS FOR FIRE MANAGEMENT IN THE GWW

Protecting the current carbon stocks of the GWW will avoid emissions of carbon dioxide (a major greenhouse gas) from continuing disturbances. In addition, land management can be undertaken that promotes the restoration of current carbon stocks where they have been degraded so that the GWW begins to approach its carbon carrying capacity. To achieve these aims, two basic options can be considered.

Do nothing

If this option is taken then: 1) the areal extent of residual woodland is likely to decrease; and 2) fire will continue to prevent the pyric successional vegetation from attaining a mature successional stage with higher biomass carbon stocks.

Regional planning and management

Ideally, the entire region could be placed under a conservation planning overlay, perhaps analogous to how the Great Barrier Reef is managed in its entirety via zoning and associated management prescriptions. The most appropriate legal framework and policy instruments to achieve this remain to be identified. Given the aims of this report, we can only suggest some elements that should be considered as part of an integrated approach to management of the GWW.

1. **Change the tenure of the remaining long-unburnt woodland, currently under vacant crown land, to nature conservation reserve.** This would make the WA Department of Environment and Conservation (WADEC) responsible for fires that are on or threaten the area. Based on information provided on the WADEC web site (viewed 24 March 2009, <http://www.dec.wa.gov.au/fire/wildfires/wildfire-response.html>), we understand that WADEC is responsible only for fires that occur on lands that it manages, and on a 'good neighbour' basis for fires occurring near to or threatening land managed by WADEC. WADEC control, however, provides no guarantee that the woodlands will not burn, as suppression activities are difficult in this remote region. For example, an intense and extensive wildfire burned through woodland in the Dundas Nature Reserve in 2002–03. Additionally, the biomass carbon stocks of pyric successional vegetation would remain static or deteriorate with continued uncontrolled fire on land outside the reserve system.

2. **Place all vacant crown lands (VCL) within the GWW into the nature conservation reserve system.** As noted above, this would not guarantee protection from fire. It would, however, give WADEC responsibility for its management. Management activities could include selected road and access track closures and the imposition of total fire ban periods during the peak fire-danger period. We note that it may be impossible for a management authority such as WADEC to provide suppression of fires in this region given: 1) the remote location; 2) the distance from water sources; and 3) other regions within Western Australia are deemed to have higher priority during the peak fire season.

3. **Establish an education campaign to reduce accidental and deliberate ignition events.** If the VCL was placed under the control of WADEC, it could take responsibility for this education campaign. It would need to be targeted at field surveyors and exploration geologists, in addition to tourists and locals who use the GWW for recreation and other activities. A combination of the above approaches is likely to confer the greatest protection of the existing biomass carbon stock as well as allowing pyric successional vegetation to mature and consequently attain increased biomass carbon, thereby enabling the GWW to approach its carbon carrying capacity.

One important question remains: assuming that it is possible for succession to proceed, how long might it take for pyric successional shrubland of eucalypt seedlings and lignotuber re-sprouts to become woodland?

7.3 ESTIMATED GROWTH RATE OF TREES IN THE GWW

There have been no studies of the growth rates of eucalypt trees in the GWW. We do, however, have two types of information that can be used to inform estimates of tree age. First, during our field survey we collected five stem discs from fallen trees. These discs were taken from several metres up the stems as the lower stems had been hollowed out by termites. Stem sections were prepared from the discs and growth rings were counted to provide estimates of stem age. The data are presented in Table 7.1. Ring counts are related to stem cross-sectional area, shown in Figure 7.5.

Table 7.1 Estimated age of stem sections from eucalypt wood samples collected in the GWW. As the GWW has a distinct annual growing season (during winter/spring), the ring counts can be equated to age increments.

Diameter under bark (cm)	Ring count estimated equivalent age (years)
29.4	169
16.3	81
9.2	66
12.1	75
17.0	78

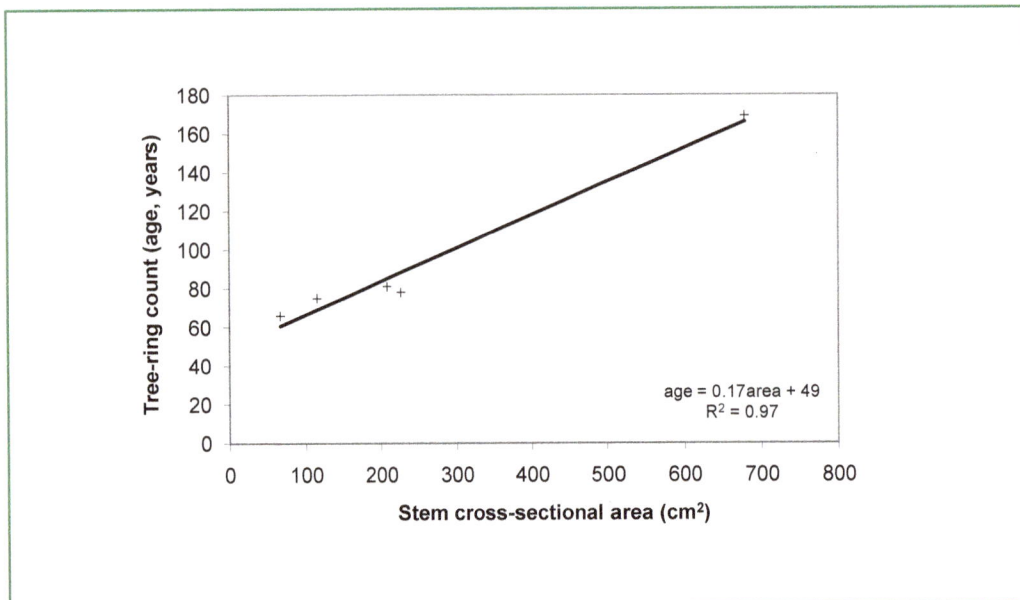

age = 0.17area + 49
$R^2 = 0.97$

Figure 7.5 Relationship between tree-ring count and stem cross-sectional area (under bark) for GWW wood samples analysed by Dr Matt Brookhouse, ANU. The equation describes the line of best fit through the data points.

The second type of information comes from our field data measurements at Field Site 7, which include diameters of re-growth and coppice re-growth following timber cutting during the period 1921–38. The estimated age of the stems is between 67 and 84 years old. Estimates of age, using the regression equation in Figure 7.5, fall mostly within this age range (Figure 7.6). Twenty per cent of these stems give a greater than expected age.

This could mean that these trees are older and were not selected for harvest by timber cutters or that the growth rate of these stems was greater than average due to greater access to resources and less competition from neighbouring stems.

By applying the stem age to cross-sectional area relationship to the five field survey sites having long-unburnt vegetation and no timber harvesting, we have estimated a likely stand age. As the distributions of estimated age were similar for all five sites, we have combined the data to produce the histogram shown in Figure 7.7. The sites contain trees of mixed age—mostly from 100 to 400 years. A lifespan of several hundred years has previously been reported for eucalypt species in other parts of Australia (Looby 2007; Ogden 1978; Wellington and Noble 1985). All five sites contained some trees that were very large and possibly very ancient.

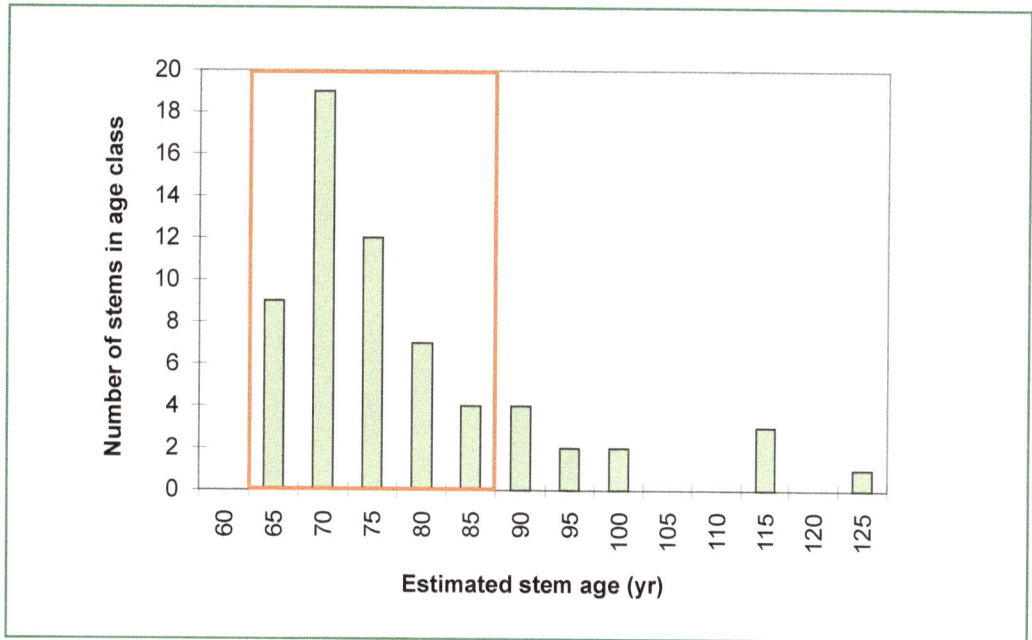

Figure 7.6 Estimated age of coppice stems measured at Field Survey Site 7 in 2005. The coppice stems have re-grown following timber cutting between 67 and 84 years before our field survey. The orange rectangle indicates those stems with an age estimate that falls within the boundaries expected from the timber-cutting data.

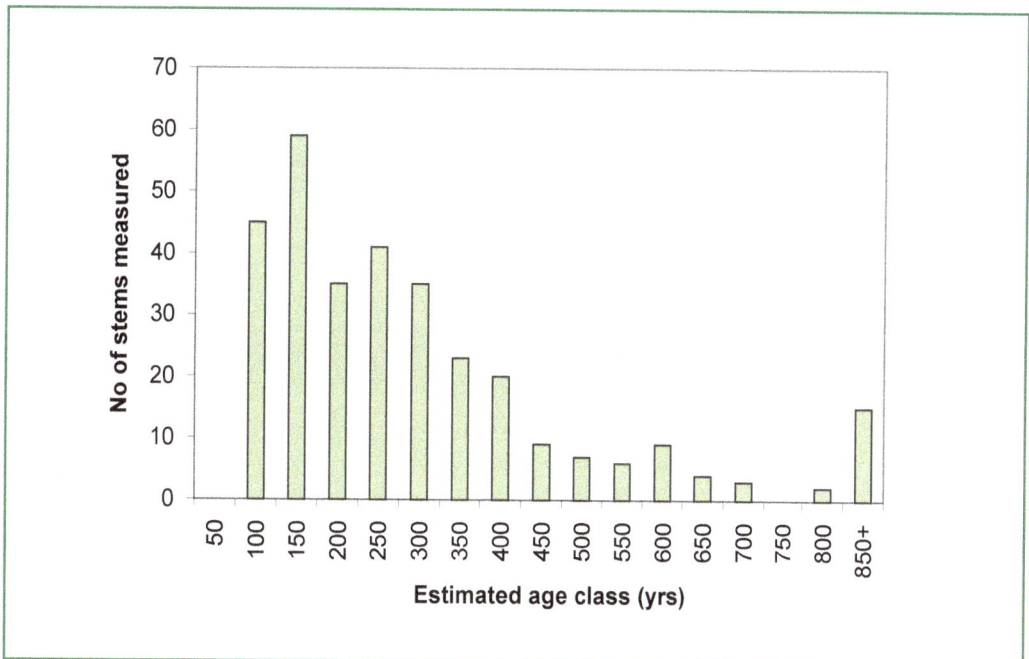

Figure 7.7 Estimated age class for trees measured at Field Survey Sites 6, 12, 13, 15 and 20. These sites showed no signs of recent disturbance from fire, timber cutting or prospecting.

This brief investigation of tree growth rates and ages suggests that it could take several centuries for the completion of pyric succession and for biomass carbon to reach maximum levels. This can be achieved only if intense and frequent fires are successfully excluded from extensive areas of the GWW.

8. CONCLUSION

In this report, we have quantified the carbon dynamics of the GWW with sufficient accuracy to support consideration of management options for actions that will help protect current carbon stocks and begin to restore the region's carbon carrying capacity. While 'no fire' could be a desirable management option, we accept that this is an unrealistic expectation and that fires will occur irrespective of the management efforts. Therefore, the most feasible goals are probably minimisation of human ignitions and partial suppression of lightning ignitions, through tenure and land-use management changes.

Implementing these management options will, however, come at a cost. Conservation management options are more likely to succeed if they can be linked to the emerging carbon market and payments for land stewardship and ecosystem services. The emerging carbon economy provides a potential source of investments for protecting and restoring green carbon stocks in the GWW. Voluntary carbon offset schemes involving changes to land management are already under way in parts of Australia (for example<http://www.carbonoffsetguide.com.au/>). Furthermore, consideration is being given to forest mitigation schemes based on payment for ecosystem services and land stewardship (Costa 2009). Therefore, it is important that state, Commonwealth and international policies and actions recognise the value of avoiding emissions from extant carbon stocks in natural ecosystems, along with the sequestration potential from managing threatening processes. Incentives are needed that will enhance carbon stocks in the GWW through ecological restoration, while avoiding perverse outcomes such as inadvertently providing incentives to clear and degrade natural vegetation ecosystems.

If regions such as the GWW are to be included in carbon market schemes, precise estimates of spatial and temporal changes in carbon stocks will be required. Consequently, there will be a need to establish a network of long-term sites where all components of the carbon stock, including soil organic carbon, are monitored.

APPENDIX

1. CALCULATION OF ABSOLUTE DENSITY OF INDIVIDUALS MAKING UP A CANOPY LAYER FROM INTENSIVE SITE DATA

For a detailed description of the methodology, see Mitchell (2007).

For each sampled canopy layer at each field site the sum of distances, D_n, is:

$$D_n = \sum_{i=1}^{16} \sum_{j=1}^{4} R_{ij}$$

in which

i is a particular transect point (1–16)

j is a quarter at a transect point (1–4)

n is the number of quarters in which a measurement is made

n_0 is the number of vacant or empty quarters

R_{ij} is the point-to-plant centre distance at point i in quarter j

The mean distance, $\bar{r}$, is:

$$\bar{r} = \frac{D_n}{n}$$

Case 1. $n_0 = 0$

The absolute density, λ, is:

$$\lambda = \frac{1}{\bar{r}^2} \quad \text{plants m}^{-2}$$

$$\lambda = \frac{10000}{\bar{r}^2} \quad \text{plants ha}^{-1}$$

Case 2. $n_0 > 0$

In this case, a correction factor (CF) is applied. First, it is necessary to calculate the ratio

$$\frac{n_0}{n + n_0}$$

Then, using the look-up table (Table 1 provided in Warde and Petranka 1981), find the CF that applies to this ratio.

The absolute density, λ, is:

$$\lambda = \frac{1}{\bar{r}^2} CF \quad \text{plants m}^{-2}$$

$$\lambda = \frac{10000}{\bar{r}^2} CF \quad \text{plants ha}^{-1}$$

2. CALCULATION OF THE BASAL AREA OF THE TREE LAYERS FROM INTENSIVE SITE DATA

For each sampled canopy layer at each field site the sum of the cross-sectional area of the measured stems, A_n, is:

$$A_n = \sum_{i=1}^{16} \sum_{j=1}^{4} \pi r^2_{ij}$$

in which r is the stem radius at 1.3 m (that is, stem dbh/2).

In the case of multi-stemmed individuals, the cross-sectional area of each stem was calculated, and the plant cross-sectional area was calculated as the sum of the areas of the individual stems making up the plant.

The mean plant stem cross-sectional area, $\overline{A}$, is:

$$\overline{A} = \frac{A_n}{n}$$

in which n is the number of quarters in which a measurement was made.

The basal area, B, is:

$$B = \overline{A}\lambda$$

Table A1 Description of spacial data layers used for model development and extrapolation.

Name or symbol	Layer description	Spatial resolution	Source
Topography and administrative			
Elevation	Ground-level elevation: Digital Elevation Model Version 3	9 second	Hutchinson et al. (2008)
Topo	Topographic position index: calculated from the 9-second digital elevation model	250 m	Gallant and Dowling (2003)
Land tenure	Australian Land Tenure 1993	1:5 000 000	Geoscience Australia (1993)
Roads	WA Roads	1:5 000 000	Geological Survey of Western Australia (2007)
Towns	Population Centres	1:1 000 000	Lawford et al. (1998)
Map sheet	1:250 000 Map Sheet Index	1:250 000	Geological Survey of Western Australia (2007)
Climate			
R_s	Solar radiation received at the surface, MJ m^{-2} yr^{-1}	1 km	Hutchinson (2005)
P	Mean annual precipitation, mm yr^{-1}	1 km	Hutchinson (2005)
T_{max} and T_{min}	Mean daily air temperature in January (Tmax) and July (Tmin), °C	1 km	Hutchinson (2005)
W	$W = P - \dfrac{R_s}{\rho L}$ mm yr^{-1}, in which ρ is the density of liquid water ($\sim$1000 kg m^{-3}) and L is the latent heat of vaporisation of water ($\sim$2.45 × 10^6 J kg^{-1} H$_2$O) $W_{3000} = W + 3000$ mm yr^{-1}. $W_{4000} = W + 4000$ mm yr^{-1}	1 km	Berry and Roderick (2002)
Vegetation			
NVEG	Natural vegetation	1:5 million	AUSLIG (1990)
PVEG	Present vegetation	1:5 million	AUSLIG (1990)
MVG	National Vegetation Information System. Australia—Present Major Vegetation Groups—NVIS Stage 1, Version 3.0	100 m	DEWHA (2005)
MODIS	Normalised Difference Vegetation Index		
MODIS 16-Day L3 Global 250m (MOD13Q1) satellite imagery	250 m	Paget and King (2008)	
Fire mapping	Landsat MSS, TM and ETM+ satellite data 1972–2002	30 m	Geoscience Australia (2010)
WA land cover	Landsat WA 2005		Geological Survey of Western Australia (2007)
Timber cutting	Timber tramlines and cutting areas in the Goldfields region, 1900–75. FD No. 1610	1:500 000	The digitised layer was created from the printed map

Table A1 Continued.

Name or symbol	Layer description	Spatial resolution	Source
Geology			
Geology	Geology of Western Australia	1:250 000	Geological Survey of Western Australia (2007)
Lithology	Surface lithology	1:500 000	Geological Survey of Western Australia (2007)
%OC	Soil carbon concentration of A and B horizons (% C)	1 km	CSIRO (2007)
$Thick_A$, $Thick_B$	Thickness of A and B horizons (m)	1 km	CSIRO (2007)
BD_A, BD_B	Soil bulk density of A and B horizons (Mg m^{-3})	1 km	CSIRO (2007)
$Thick_{AB}$	$Thick_{AB} = Thick_A + Thick_B$ (m)	1 km	
SOC_A	$SOC_A = Thick_A\ BD_A\ \dfrac{\%OC}{100}$ (kg C m^{-2} of ground surface)	1 km	
SOC_B	$SOC_B = Thick_B\ BD_B\ \dfrac{\%OC}{100}$ (kg C m^{-2} of ground surface)	1 km	

Table A2 Summary of explanatory notes for the 1:250 000 series Geological Map Sheets covering the GWW relating to road access, vegetation cover and water. Quotes are italicised.

Map Sheet Explanatory notes, author and year	Accessibility for survey	Vegetation notes	Water notes
Jackson SH5012 Chin and Smith (1981)	*Farms occupy a small portion of the southwestern part of JACKSON, and there has been intermittent grazing on parts of the uncleared remainder. Access in the northern portion of JACKSON is poor due to the paucity of roads and tracks and thick vegetation.*	*The vegetation...is chiefly characterized by open sclerophyll woodland on the red soil slopes and valleys, by thick acacia scrub on sandplains and ironstone ridges, and by saltbush with acacia scrub adjoining saline drainages and lakes.*	*Prospects for potable shallow groundwater are poor.*
Southern Cross SH5016 Gee (1979)	*This sheet is crossed by Great Eastern Highway, Perth to Kalgoorlie railway and Eastern Goldfields water-supply pipeline.* *The eastern part [that is, the GWW] is mostly uncleared vacant crown land, which has been used for mining and timber gathering.*	*This eastern part supports heath on sand plains, woodlands in valleys, and halophytes around salt lakes.*	No comment
Hyden SI5004 Chin et al. (1984)	*East of the vermin-proof fence, vehicle access is generally poor and is restricted to a few tracks and overgrown cut-lines. However, throughout the Forrestania Greenstone Belt, numerous tracks and cut-lines have been constructed by mining companies during exploration for base metals (particularly nickel), and these provide reasonable access for four-wheel drive vehicles.*	*Cites Beard (1972)* *Scrub heath occupies the sandplain ridges; mallee in the middle slope, covering the largest portion of the area; and sclerophyll woodland the lowest ground. In low-lying saline areas, woodland is replaced by Ti-tree scrub while samphire is predominant in hypersaline areas. The heavy soil (Ae unit) of the Forrestania Greenstone Belt is mostly covered by sclerophyll woodland. Cites Beard (1972)*	*Present day drainage is internal into salt lakes.* *No large quantities of good quality underground water have yet been located. However, small quantities of good quality water have been obtained from the deepest Quaternary and Tertiary sands on slopes flanking large granite outcrops. Groundwater salinities are higher at greater distances from the recharge area and are mostly too high for agricultural use.*
Kalgoorlie SH5109 Kriewaldt (1967)	No comment	No comment	*At present there are some 20 bores and wells within the Kalgoorlie Sheet area that are being used for watering sheep. The prospects of getting good quality underground water in large supplies are very small.*

Table A2 Continued.

Map Sheet Explanatory notes, author and year	Accessibility for survey	Vegetation notes	Water notes
Boorabbin SH5113 Hunter (1989) Kern (1995)	Mineral exploration and mining co-exist with primary production in the northeast corner of BOORABBIN...land west of the Coolgardie-Norseman Highway have reverted to Crown land. The remainder is vacant Crown land, which was extensively exploited for timber in the first half of this century. Access throughout BOORABBIN is generally poor. Since 1965 the region has been intensively explored, first for nickel and other base-metals and, more recently, for gold.	There is a close correlation between photo-interpreted geology and the six 'plant formations' delineated by Beard (1976), since the success of floral species is strongly controlled by soil type.	Good water catchment areas around some large granite outcrops.
Lake Johnston SI5101 Gower and Bunting (1976)	Access within sheet is generally poor except for areas of mineral exploration activity in the south-eastern corner and Bremer Range. Off-track driving is hampered by dense scrub.	There is a close correlation between the vegetation units and the Cainozoic soil units mapped during the present survey. Both were mapped using aerial photo-patterns that are largely dependent on vegetation.	Poor prospects for groundwater due to low rainfall and high evaporation. Small supplies of stock-quality water possibly in catchments of large granite outcrops. Possible fresh water beneath larger sand dunes.
Ravensthorpe SI5105 Thom et al. (1977)	The undeveloped northern third of the sheet is easily accessed by numerous cut-lines and tracks.	Much of the area is covered by scrubland heath and woodlands of small eucalypts; in general large trees are rare.	Within the GWW, no developed groundwater supplies except for three abandoned soaks.
Kurnalpi SH5110 Williams (1970)	Graded roads link the scattered station homesteads to Kalgoorlie. Numerous pastoral station tracks provide access to most of the area.	The amount of vegetation belies the semi-arid nature of the climate. Open eucalypt woodlands which consist mainly of salmon gums, gimlet, and mallee, interspersed with saltbush and bluebush, cover the flat undulating country to the south. Vegetation density increases towards the southern margin of the sheet area.	No comment

100

Table A2 Continued.

Map Sheet Explanatory notes, author and year	Accessibility for survey	Vegetation notes	Water notes
Kurnalpi SH5110 Williams (1970)	Graded roads link the scattered station homesteads to Kalgoorlie. Numerous pastoral station tracks provide access to most of the area.	The amount of vegetation belies the semi-arid nature of the climate. Open eucalypt woodlands which consist mainly of salmon gums, gimlet, and mallee, interspersed with saltbush and bluebush, cover the flat undulating country to the south. Vegetation density increases towards the southern margin of the sheet area.	No comment
Widgiemooltha SH5114 Sofoulis (1966)	No comment	No comment	With the general low relief and low annual rainfall over the region, most of the groundwater available is saline and not suited for stock use. Some good water from large bare granite rocks. Some domestic supplies are obtained from roof catchment. Goldfields Water Supply pipeline serves Widgiemooltha township, other small sidings, mining centres and pastoral properties adjacent to Norseman–Coolgardie road.
Norseman SI5102 Doepel (1973)	The country is easily accessible by road and track near Norseman, in the agricultural area near Salmon Gums, and in the pastoral area of Fraser Range. However, over the greater portion of the Sheet area there are only a few tracks. Travel away from the tracks is possible by four-wheel drive vehicle, but is hampered by dense scrub.	The vegetation ranges from an open eucalypt woodland to a dense scrub (especially in the south). It is composed of various species of Eucalyptus, Acacia, Casuarina, Melaleuca, and saltbush (Atriplex) and bluebush (Kochia). Open grassland with spinifex (Triodia) occurs on parts of the Jimberlana Dyke and the Fraser Range.	
Esperance SI5106 Morgan and Peers (1973)	No comment	No comment	No comment

Table A2 Continued.

Map Sheet Explanatory notes, author and year	Accessibility for survey	Vegetation notes	Water notes
Cundeelee SH5111 Bunting and van de Graaff (1977)	Tracks within the sheet are restricted to the western third and eastern margin.	Southwest of Ponton Creek eucalypt woodlands predominate, being gradually replaced by mulga towards the northwest.	The prospects of finding water of stock quality or better on the Cunderlee Sheet are very small.
Zanthus SH5115 Doepel and Lowry (1970b)	Few tracks over most of the sheet area. Off-track driving is hampered by dense scrub, especially on western part of the sheet.	Open myall (Acacia spp.) scrub on Bunda Plateau. Precambrian geology supports open eucalypt woodland to dense scrub dominated by mallee.	Only two bores near Uraryie Rock (of 20 drilled) yielded good quantities of stock water. Others were dry or saline.
Balladonia SI5103 Doepel and Lowry (1970a)	The main access is from the Eyre Highway. There are only a few tracks over the greater portion of the sheet area. Dense scrub hampers off-track four-wheel-drive travel.	South of the highway, and on the western part of the sheet there is dense scrub 10 to 20 feet high which is dominated by mallee-type eucalypt.	Pastoralists rely on small dams, excavated in colluvium flanking large granitic outcrops, for water storage. There are no bores or wells in the western portion of the Sheet area, but some underground water is likely to be present in the alluviated valleys.
Malcolm–Cape Arid SI5107 Lowry and Doepel (1974)	No comment	No comment	No comment
Culver SI5104 Lowry (1970)	The main access is from the Eyre Highway. North of the highway there are numerous vehicle tracks across open plain. South of the highway vehicle access is restricted to a few stony overgrown tracks through dense mallee scrub.	Vegetation consists mainly of blue-bush (Kochia sedifolia), salt bush (Atriplex spp.), and grasses in the northeastern part of the Sheet; mallee-type eucalypt scrub 5 to 20 feet high near the coast; and myall scrub 10 to 20 feet high in other areas.	In a few areas there could be suitable groundwater for sheep (salinities from 12 000 to 20 000 ppm) but not for cattle, as they require salinity less than 10 000 ppm.

Table A3 Relationship between vegetation cover and soil type in the GWW.

Soil group	Low shrubland	Tall shrubland	Mallee	Woodlands
Aeolian loams				Callitris columellaris, Casuarina cristata, Casuarina obesa
Aeolian sands	Atriplex vesicaria	Acacia jennerae; Acacia ligulata	E. aff. occidentalis; E. cylindrocarpa; E. eremophila; E. falcata; E. foecunda; E. gracilis; E. incrassata; E. leptophylla; E. mixed; E. oleosa; E. oleosa over Triodia scariosa; E. pileata	Callitris columellaris; E. aff. foecunda; E. eremophila; E. foecunda; E. gracilis; E. platycorys; E. salicola
Alluvial	Tecticornia verrucosa			
Alluvium		Melaleuca aff. preissiana		E. georgei
Calcareous earths	Cratystylis conocephala		E. cylindrocarpa; E. mixed mallee over Melaleuca; E. transcontinentalis	E. dundasii; E. mixed; E. oleosa; E. salmonophloia
Deep calcareous earths	Maireana sedifolia		E. cylindriflora; E. cylindrocarpa; E. gracilis; E. incrassata; E. scyphocalyx; E. spathulata ssp. grandiflora	E. (mixed); E. campaspe; E. clelandii – E. lesouefii; E. diptera; E. dundasii; E. flocktoniae; E. lesouefii; E. longicornis; E. melanoxylon; E. myriadenia; E. salmonophloia over Maireana sedifolia; E. salubris; E. sheathiana; E. transcontinentalis
Deep sands	Mixed Grevillea spp.	Acacia beauverdiana; Allocasuarina corniculata; Callitris preissii; Callitris preissii ssp. verrucosa; Grevillea eriostachya ssp. Excelsior; Grevillea excelsior; Melaleuca uncinata	E. aff. decipiens; E. burracoppinensis; E. hypochlamydea; E. leptopoda; E. platycorys; E. tetragona	
Fine loamy sands			E. transcontinentalis	

103

Table A3 Continued.

Soil group	Low shrubland	Tall shrubland	Mallee	Woodlands
Fine sandy clays			E. diptera	Acacia acuminata
Granitic loam		Thryptomene australis	E. gracilis	
Granitic sands				
Granitic soils		Acacia acuminata; Acacia quadrimarginea; Acacia sessilispica; Acacia sp. (KRN 7568); Allocasuarina campestris ssp. campestris; Leptospermum erubescens; Melaleuca uncinata	E. grossa; E. loxophleba	Acacia lasiocalyx; Allocasuarina huegeliana
Granitic light clay		Acacia		
Gravelly sands	Hakea cf. falcata	Acacia beauverdiana; Acacia signata; Allocasuarina acutivalvis; Allocasuarina campestris ssp. campestris	E. scyphocalyx; 1	
Gritty loams		Hakea pendula		E. capillosa
Gritty sands		Allocasuarina acutivalvis		
Gritty soils		Baeckea sp. (KRN 7010)		
Gypsum soils				E. fraseri
Light clays	Halosarcia			
Loamy clay sands			E. effusa	
Loamy clays				E. creta
Loamy sands		Callitris canescens; Melaleuca uncinata		E. melanoxylon
Meta-granitic soils			E. loxophleba	
Red cracking clays	Cratystylis subspinescens			E. salubris

104

Table A3 Continued.

Soil group	Low shrubland	Tall shrubland		Mallee	Woodlands
Red sands			Acacia quadrimarginea	E. oleosa	
Saline soils	Halosarcia spp.	Melaleuca uncinata		Melaleuca uncinata	
Shallow calcareous earths				E. griffithsii	Casuarina cristata; E. corrugata; E. fraseri; E. oleosa; E. oleosa – Casuarina cristata; E. sp. (KRN5603); E. stricklandii; E. torquata
Shallow loamy sands				E. oleosa	E. oleosa; E. transcontinentalis
Shallow sands		Melaleuca		E. aff. occidentalis; E. pileata; E. redunca; E. transcontinentalis	
Shallow sandy clays				E. celastoides var. virella	
Sub-saline soils	Halosarcia spp.; Atriplex vesicaria ssp. variabilis	Dodonaea angustissima; Eremophila miniata; Melaleuca aff. cuticularis			E. sp. (KRN 9710)

Eucalyptus is abbreviated to E.

Compiled from field site data published in the records of the Western Australian Museum (Newby et al. 1984).

Table A4 Relationship between plant formation and soil type in the areas covered by the Boorabbin and Lake Johnston 1:250 000 map sheets.

Soil Type	Vegetation Type					
	1. Scrub heath	**2. Broombush thicket**	**3. Rock pavement vegetation**	**4. Mallee and marlock**	**5. Sclerophyll woodland (trees incl. tree mallee)**	**6. Halophytes**
	Mostly ~1 m tall; very frequently burnt	Up to 3 or 4 m tall; intergrades with scrub heath		~3–5 m tall, occasionally 6–8 m; frequently burnt	Mostly 12–20 m tall	
Leached sands	Partly open canopy including Proteaceae and Myrtaceae (but no *Eucalyptus* spp.)					
Shallow sandy soil over laterite, ironstone and gravel, or unweathered granite		Dense shrub assemblage comprising mostly *Casuarina, Acacia* and *Melaleuca* spp.				
Soil-less granite outcrops			Lichen and moss, with aquatic plants in pools and shrubs in crevices or soil patches			
Highly saline depressions						Chenopod shrubs near salt lakes
Granite—residual or eluvial						
(i) Red loam	Understorey shrub layer incl. *Alyxia, Conosperma, Daviesia, Eremophila, Grevillea, Scaevola, Westringia* spp.	Understorey tall shrub layer incl. *Dodonaea, Melaleuca, Santalum* spp.			*Eucalyptus transcontinentalis* – *E. flocktoniae* association. Other trees incl. *E. corrugata, E. salubris, E. melanoxylon*	
(ii) Leached soils	Understorey of mixed heath shrubs					

Soil Type	Vegetation Type				
Hummock grass on red sands	Understorey shrub layer incl. *Melaleuca*		*E. eremophila* in association with one or more of: *E. redunca, E. incrassata, E. pileata, E. leptophylla, E. flocktoniae, E. loxophleba*	Saltbush understorey on alkaline soils	
Granite—alluvial					
(i) Light loam	Sparse understorey can include *Eremophila, Acacia, Davesia*	Tall understorey shrub layer of *Melaleuca*		*E. salmonophloia*	
(ii) Clay-loam	Sparse understorey can include *Eremophila, Acacia, Davesia*	Tall understorey shrub layer of *Melaleuca*		*E. salmonophloia – E. longicornis*	
(iii) Stiff clay	Sparse understorey can include *Eremophila, Acacia, Davesia*	Tall understorey shrub layer of *Melaleuca*		*E. salmonophloia – E. salubris*	
(iv) Sandridges				*E. aff. striaticalyx-E. leptophylla* association with occasional *E. salmonophloia*	Chenopod shrub understorey
Greenstone— residual or eluvial					

Table A4 Continued.

Soil Type	Vegetation Type			
(i) Rocky ridges	Open understorey incl. Cassia	Open understorey incl. Eremophila, Dodonaea, Acacia spp.	E. torquate – E. lesoeufii co-dominant. Other trees include E. corrugata, E. clelandii, E. campaspe and Casuarina cristate	Atriplex (old man saltbush)
(ii) Deep soil	Understorey shrub layer incl. Alyxia, Connesperma, Davesia, Eremophila, Grevillea, Scaevola, Westringia spp.	Understorey tall shrub layer incl. Dodonaea, Melaleuca, Santalum spp.	E. dundasii – E. longicornis; E. lesoeufii – E. oleosa	
Greenstone—alluvial				
Soils of high base status	Sparse understorey can include Eremophila, Acacia, Davesia	Tall understorey shrub layer of Melaleuca	E. salmonophloia – E. melanoxylon	

Species found in vegetation types 1, 2 and 6 can make up the understorey of types 4 and 5 where soil type is appropriate.
Data compiled from Beard (1968).

Table A5 Summary of data obtained from field survey of 21 sites in the GWW during 2005–07.

Site number	Survey ID	Canopy layer	Number of observations	Absolute density (plants ha^{-1})	Basal area (m^2 ha^{-1})	Mean canopy height (m)	Stdev canopy height (m)
1	2005_01	T	52	37	4.90	16.95	3.90
1	2005_01	S	61	96	no data	4.33	2.43
1	2005_01	C	64	759	no data	0.60	0.23
2	2005_02	T1	20	10	1.48	11.99	3.52
2	2005_02	T2	19	16	no data	2.73	0.92
2	2005_02	Y	32	330	no data	0.94	0.34
2	2005_02	S	42	62	no data	2.92	0.67
2	2005_02	Z	64	2 433	no data	0.55	0.43
3	2005_03	Z1	28	22	no data	1.26	0.24
3	2005_03	Z2	64	1 443	no data	0.66	0.11
3	2005_03	Z3	64	26 120	no data	0.31	0.09
4	2005_04	Z1	64	136	no data	1.35	0.43
4	2005_04	Z2	64	3 999	no data	0.61	0.10
4	2005_04	Z3	64	112 513	no data	0.32	0.08
5	2005_05	T	49	28	5.74	16.14	2.32
5	2005_05	M	53	96	2.66	4.85	2.71
5	2005_05	S	64	295	no data	1.79	0.73
5	2005_05	Z1	64	1 069	no data	0.75	0.14
5	2005_05	Z2	64	2 253	no data	0.29	0.09
6	2005_06	T	58	45	6.04	12.79	3.72
6	2005_06	C	64	1 682	no data	0.78	0.19
6	2005_06	Z1	52	49	no data	1.68	0.85
6	2005_06	Z2	64	1 988	no data	0.35	0.10
7	2005_07	T1	63	109	8.56	10.27	2.92
7	2005_07	T2	31	26	no data	6.66	1.71
7	2005_07	Z1	64	478	no data	1.53	0.39

Table A5 continued.

Site number	Survey ID	Canopy layer	Number of observations	Absolute density (plants ha⁻¹)	Basal area (m² ha⁻¹)	Mean canopy height (m)	Stdev canopy height (m)
7	2005_07	Z2	64	932	no data	0.71	0.20
7	2005_07	Z3	64	2 058	no data	0.34	0.09
8	2005_08	Y	52	50	no data	3.57	1.27
8	2005_08	Z1	63	304	no data	2.66	0.62
8	2005_08	Z2	64	2 156	no data	1.21	0.34
9	2005_09	T1	39	21	2.94	12.84	2.53
9	2005_09	T2	42	27	1.39	8.61	1.46
9	2005_09	Z1	64	425	no data	1.60	0.74
9	2005_09	Z2	64	719	no data	0.73	0.15
10	2005_10	Y	27	38	no data	3.81	1.07
10	2005_10	Z1	64	2 252	no data	1.83	0.48
10	2005_10	Z2	64	19 083	no data	0.71	0.17
11	2006_01	T	57	124	1.42	7.31	2.97
11	2006_01	Y	13	12	no data	4.83	3.99
11	2006_01	S	52	358	no data	1.78	0.40
11	2006_01	Z	62	523	no data	0.70	0.22
12	2006_02	T	58	67	6.64	13.22	3.58
12	2006_02	M	53	58	0.94	4.59	2.33
12	2006_02	S	59	93	no data	2.63	1.06
12	2006_02	Z	64	853	no data	1.02	0.28
13	2006_03	T1	39	19	11.54	14.79	4.17
13	2006_03	T2	10	3	0.30	8.57	1.25
13	2006_03	T3	20	11	0.05	3.76	1.48
13	2006_03	S	62	215	no data	3.37	1.78
13	2006_03	Z	64	1 887	no data	0.84	0.23
14	2006_04	T1	38	19	4.16	16.92	4.35
14	2006_04	T2	39	19	0.77	12.52	2.85
14	2006_04	T3	14	6	0.00	3.66	1.75
14	2006_04	M	31	28	0.81	7.65	1.63

Table A5 continued.

Site number	Survey ID	Canopy layer	Number of observations	Absolute density (plants ha⁻¹)	Basal area (m² ha⁻¹)	Mean canopy height (m)	Stdev canopy height (m)
14	2006_04	S	64	337	no data	2.49	0.93
14	2006_04	Z	64	1 384	no data	0.89	0.25
15	2006_05	T1	50	29	4.80	11.80	2.71
15	2006_05	T2	28	17	0.58	9.39	1.79
15	2006_05	T3	10	5	0.02	4.48	2.60
15	2006_05	S1	45	29	no data	2.44	0.90
15	2006_05	S2	64	328	no data	0.84	0.29
16	2006_06	M	7	4	0.07	6.26	1.08
16	2006_06	S1	64	364	no data	3.57	0.80
16	2006_06	S2	64	9 876	no data	0.93	0.21
17	2006_07	M	18	8	0.33	7.33	3.29
17	2006_07	S1	64	470	no data	3.24	0.88
17	2006_07	S2	64	7 707	no data	0.96	0.31
18	2007_01	T/Y	60	330	3.22	6.50	1.45
18	2007_01	S	64	2 436	no data	2.77	0.64
18	2007_01	Z	64	1 197	no data	0.99	0.31
19	2007_02	T1	46	26	4.55	12.81	2.23
19	2007_02	T2	51	7	0.24	10.43	2.21
19	2007_02	T3	53	11	0.51	4.95	0.99
19	2007_02	S	64	116	no data	2.70	0.66
19	2007_02	C/Z	64	564	no data	0.86	0.20
20	2007_03	T1	60	54	9.11	16.45	2.83
20	2007_03	T2	49	35	1.26	11.89	2.26
21	2007_04	Y	50	124	no data	2.10	0.68
21	2007_04	Z	64	12 541	no data	0.79	0.21

T = monopodial tree canopy layer (numbered T1, T2, T3 when there is more than one tree canopy layer). M = mallee tree layer. S = tall shrub layer (incl. Acacia, Melaleuca, Santalum). C = saltbush/ bluebush low shrub layer. Y = mallee shrub layer. Z = sclerophyllous (heath) shrub layer .
Basal area estimates are available only for the tree layers. The maximum possible number of observations for a layer at a site is 64 (16 point centres × 4 quarters).

Table A6.1 Biomass and allometric regression equations for Queensland woodland tree/mallee species.

Description	n	DBH range (cm)	a	b	R^2	RSD
Mallee (Eucalyptus socialis/E. dumosa combined) (Burrows et al. 1976) (33°53'S, 146°30'E) LnY = a + bLnX in which X = circumference at 30 cm Total mass (kg)	29	2.5–55.2	–4.1671	2.2620	0.98	0.213
Mallee re-growth (Eucalyptus socialis/E. gracilis/E. leptophylla combined) (Burrows et al. 1976) (33°53'S, 146°30'E) LnY = a + bLnX in which X = circumference at 30 cm. Total mass (kg)	30	1.8–23.2	1.9943	2.4780	0.99	0.165
Callitris glaucophylla trees, Qld (Burrows et al. 2001) $y = a + be^{(-x/c)}$ x = circumference at 30 cm (cm) y = total above-ground mass (kg) c = –56.704	20		–110.512	81.962	0.976	37.393
Callitris glaucophylla trees, Qld (Burrows et al. 2001) $y = a + be^{(-x/c)}$ x = circumference at 130 cm (cm) y = total above-ground mass (kg) c = –50.193	20 0.989	25.277	–110.195	83.251	0.989	25.277

Source: Eamus et al. (2000:Appendix).

Table A6.2 Biomass and allometric regression equations for Queensland woodland shrub species.

Description	n	Height range (m)	a	b	R^2	RSD*
Acacia aneura (mulga) shrubs <4.5 m high (Harrington 1979)						
$LnY = a + bLnX$						
in which X = shoot height (m)						
Leaf (kg)	19	0.2–3.5	-2.589	2.116	0.941	0.476
Wood (kg)	19	0.2–3.5	-1.736	2.404	0.941	0.5448
Acacia harpophylla (brigalow) (Scanlan 1991)						
$Y = e^{(a+bLnX)}$						
in which X = shoot height (cm)						
Total above-ground biomass (g)	29	0.6–3.2	-4.303	2.150	0.86	0.558
Cassia nemophila shrubs <4.5 m high (Harrington 1979)						
$LnY = a + bLnX$						
in which X = shoot height (m)						
Leaf (kg)	19	0.6–2.0	-1.867	2.286	0.865	0.392
Wood (kg)	19	0.6–2.0	-1.310	3.297	0.884	0.494
Dodonaea viscosa shrubs <4.5 m high (Harrington 1979)						
$LnY = a + bLnX$						
in which X = shoot height (m)						
Leaf (kg)	40	0.2–2.0	-3.940	2.492	0.792	0.9823
Wood (kg)	40	0.2–2.0	-3.275	3.380	0.884	0.922

* Residual Standard Deviation

Table A6.2 continued.

Description	n	Height range	a	b	R²	RSD
Eremophila bowmanii shrubs <4.5 m high (Harrington 1979) LnY = a + bLnX in which X = shoot height (m)						
Leaf (kg)	18	0.2–1.8	−3.236	2.586	0.903	0.5448
Wood (kg)	18	0.2–1.8	−0.259	3.522	0.941	0.5448
Eremophila mitchelli shrubs <4.5 m high (Harrington 1979) LnY = a + bLnX in which X = shoot height (m)						
Leaf (kg)	18	0.6–5.0	−2.612	2.532	0.828	0.6914
Wood (kg)	18	0.6–5.0	−1.790	3.002	0.922	0.545
Myoporum deserti shrubs <4.5 m high (Harrington 1979) LnY = a + bLnX in which X = shoot height (m)						
Leaf (kg)	17	0.2–2.0	−1.535	2.449	0.846	0.6914
Wood (kg)	17	0.2–2.0	−0.998	3.030	0.923	0.643

Source: Eamus et al. (2000:Appendix).

Vegetation type	Dominant species	State	Site	Equation form	Y	X	a	b	SE(a)	SE(b)	R²(adj)	P value	n	Source
Heathland	Hakea and Banksia spp.	WA	30°23'S, 115°30'E	Y = a + bX	kg x 10⁴ ha⁻¹	% cover	0.299	0.026	0.164	0.005	0.850	0.0002	115	Rullo (1981)
Heathland	Banksia ornata only	Vic.	36°36'S, 141°20'E	Y = a + bX	g m⁻²	Canopy diameter (m)	−13 578	219			0.897		11	Ata (1996)
Heathland	Other woody shrubs (other than B. ornata)	Vic.	36°36'S, 141°20'E	Y = a + bX	g m⁻²	Canopy diameter (m)	−308	17			0.669		88	Ata (1996)
Heathland	Hakea and other species	WA	30°23'S, 115°30'E	Y = a + bX	t ha⁻¹	% cover	2.99	0.25			0.85		115	Delfs et al. (1997)
Shrubland	Acacia and Hibbertia spp.	SA	34°36'S, 138°51'E	Y = a + bX	kg ha⁻¹	% cover	−0.095	36.26	9.169	1.377	0.921	<0.0001	60	Mattiske (1975)
Open woodlands (mallee)	Eucalyptus socialis	SA	34°21'S, 139°37'E	Y = a + bX	kg	Canopy height	−0.195	0.323			0.975	<0.0001	7	Orell (n.d.)
Open woodlands (mallee)	Eucalyptus socialis/oleosa	SA	34°21'S, 139°37'E	Y = a + bX	tonnes	Mean basal stem diameter (cm)	−8.778	3.607	7.508	0.862	0.734	0.0086	7	Neagle (n.d.)
Open woodlands	Acacia acuminata and E. loxophleba	WA	31°24'S, 117°45'E	Y = a + b (lnX)	Height (m)	Canopy area (m2)	2.701	1.406	0.770	0.313	0.827		5	Van Schagen (1989)

Source: Grierson et al. (2000:Table 2, p. 5; unpublished information).

Table A6.4 Allometric relationships (lnBiomass = a + b lnDiameter at 30 cm) between above-ground biomass and coarse root biomass with stem diameter for *Eucalyptus populnea* woodlands along a rainfall gradient.

Location	Site	Rainfall (mm)	a	b	SE(a)	SE(b)	Mean square error	n	Diameter range (cm)
Oakvale	30.92°S, 46.50°E	367	−1.370	2.079	0.443	0.114	0.0391	9	15.0–107.9
Roma	25.75°S, 148.41°E	602	−2.824	2.581	0.147	0.056	0.0454	10	1.6–51.5
Rockhampton	23.17°S, 50.56°E	1 103	−2.388	2.411	0.327	0.101	0.0967	9	3.2–71.1

Source: Zerihun et al. (2006).

Table A7 Distribution, ecology and wood properties of species used by Justin Jonson to derive allometric equations*.

Species name	Common name	Distribution	Ecology	Source (see below)	[1]Basic density kg m^{-3})
Eucalyptus platypus	Moort	Albany to Esperance	Southern coastal and sub-coastal plains (B&K). Sandy soils, loamy clay, laterite. Plains, hilly and rocky country (Florabase, <http://florabase.calm.wa.gov.au/browse/profile/5643)	2	
E. falcata	Silver mallee	In a triangle bounded by Albany-Perth-Esperance	South-west, outside of wetter areas (B&K). Sand over laterite, often with gravel, limestone. Sandplains, breakaways, slight rises, hilltops, valleys, disturbed land, road verges (Florabase)	2	
E. flocktoniae	Merrit	Widespread in wheatbelt and Goldfields	Sandy loam or clay, laterite (Florabase)	2	1 145, 1 074 (820)
E. captiosa		Wheatbelt, Albany to Hyden	White or yellow sand. Sandplains, lateritic rises (Florabase)	2	
E. annulata	Open-fruited mallee	Southern wheatbelt and southern coastal areas from near Albany to Balladonia	Clay, sandy clay, clay loam (Florabase)	2	959 750
E. occidentalis	Swamp yate	Southern wheatbelt and sub-coastal areas	Usually confined to wet, clayey depressions (B&K), Sandy or clayey soils. Alluvial flats, low-lying wet areas, around salt lakes, hills (Florabase)	2	776
Allocasuarina huegeliana		Occurs from Murchison River and Mingenew south to the south coast and east to Newman Rock, west of Balladonia, WA	Associated with granite	3	885 (700)
Acacia saligna		Wheatbelt and southern Goldfields		4	

* The basic density values are estimated for air-dry (12 per cent moisture content) wood.

Sources: 1. Ilic et al. (2000); 2. Brooker and Kleinig (2001); 3. Flora of Australia Online (<http://www.environment.gov.au/biodiversity/abrs/online-resources/flora/main/>); 4. Florabase: The Western Australian Flora (<http://florabase.calm.wa.gov.au/>).

REFERENCES

Australian Surveying and Land Information Group (AUSLIG) 1990, *Atlas of Australian Resources. Volume 6: Vegetation,* (Third series), Australian Surveying and Land Information Group, Department of Administrative Services, Canberra. [Present and natural vegetation 1:5 million digitised maps are available for download, <https://www.ga.gov.au/>]

Beard, J. S. 1968, 'The vegetation of the Boorabbin and Lake Johnston areas, Western Australia', *Proceedings of the Linnean Society of New South Wales,* vol. 93, pp. 239–68.

Beard, J. S. 1972, *Vegetation Survey of Western Australia—The vegetation of the Hyden area, map and explanatory memoir,* Vegmap Publications, Sydney.

Beard, J. S. 1976, *The Vegetation of the Boorabbin and Lake Johnston Areas: Vegetation survey of Western Australia, 1:250,000 map and explanatory memoir,* Vegmap Publications, Western Australia.

Bejan, A., Lorente, S. and Lee, J. 2008, 'Unifying constructal theory of tree roots, canopies and forests', *Journal of Theoretical Biology,* vol. 254, pp. 529–40.

Berry, S. L. and Roderick, M. L. 2002, 'Estimating mixtures of leaf functional types using continental-scale satellite and climatic data', *Global Ecology and Biogeography,* vol. 11, pp. 23–40.

Berry, S. L. and Roderick, M. L. 2006, 'Changing Australian vegetation from 1788 to 1988: effects of CO_2 and land use change', *Australian Journal of Botany,* vol. 54, pp. 325–8.

Berry, S., Mackey, B. and Brown, T. 2007, 'Potential applications of remotely sensed vegetation greenness to habitat analysis and the conservation of dispersive fauna', *Pacific Conservation Biology,* vol. 13, pp. 120–7.

Bianchi, P., Bridge, P. and Tovey, R. 2008, *Early Woodlines of the Goldfields—The untold story of the woodlines to World War II,* (Second edition), Hesperian Press, Carlisle, WA.

Boland, D. J., Brooker, M. I. H., Chippendale, G. M., Hall, N., Hyland, B. P. M., Johnston, R. D., Kleinig, D. A. and Turner. J. D. 1984, *Forest Trees of Australia,* CSIRO, Melbourne.

Bond, W. J. and Keeley, J. E. 2005, 'Fire as a global "herbivore": the ecology and evolution of flammable ecosystems', *Trends in Ecology and Evolution,* vol. 20, pp. 387–94.

Bonham, C. D. 1989, *Measurements for Terrestrial Vegetation,* John Wiley & Sons, New York, pp. 159–65.

Bradby, K. 2008, 'Box 1.2—drawing a line in the sand', in A. Watson, S. Judd, J. Watson, A. Lam and D. Mackenzie, *The Extraordinary Nature of the Great Western Woodlands,* The Wilderness Society of Western Australia Inc., WA.

Brady, N. C. and Weil, R. R. 2002, *The Nature and Properties of Soils,* (Thirteenth edition), Prentice Hall, Upper Saddle River, NJ.

Brooker, M. I. H. and Kleinig, D. A. 2001, *Field Guide to Eucalypts. South-western and southern Australia. Volume 2*, (Second edition), Bloomings Books, Melbourne.

Bryant, C. 2008, *Understanding bushfire: trends in deliberate vegetation fires in Australia*, Australian Institute of Criminology Technical and Background Paper No. 27, Australian Government, Canberra.

Bunting, J. A. and van de Graaff, W. J. E. 1977, *Cundeelee, Western Australia, 1:250 000 geological series explanatory notes*, Sheet SH51-11, Explanatory notes, Bureau of Mineral Resources, Australia.

Burrows, W. H., 1976, *Aspects of nutrient cycling in semi-arid mallee and mulga communities*. PhD Thesis, Australian National University, Canberra.

Burrows, W. H., Henry, B. K., Back, P. V., Hoffmann, M. B., Tait, L. J., Anderson, E. R., Menke, N., Danaher, T., Carter, J. O. and McKeon, G. M. 2002, 'Growth and carbon stock change in eucalypt woodlands in northeast Australia: ecological and greenhouse sink implications', *Global Change Biology*, vol. 8, pp. 769–84.

Burrows, W. H., Hoffmann, M. B., Compton, J. F. and Back, P. V. 2001, *Allometric relationships and community biomass stocks in white cypress pine (Callitris glaucophylla) and associated eucalypts of the Carnarvon area—south central Queensland*, NCAS Technical Report No. 33, Australian Greenhouse Office, Canberra.

Burrows, W. H., Hoffmann, M. B., Compton, J. F., Back, P. V. and Tait, L. J. 2000, 'Allometric relationships and community biomass estimates for some dominant eucalypts in central Queensland woodlands', *Australian Journal of Botany*, vol. 48, pp. 707–14.

Chen, X., Hutley, L. B. and Eamus, D. 2003, 'Carbon balance of a tropical savanna of northern Australia', *Oecologia*, vol. 137, pp. 405–16.

Chin, R. J. and Smith, R. A. 1981, *Jackson, Western Australia, 1:250 000 geological series explanatory notes*, Sheet SH/50-12, Geological Survey of Western Australia, Record 1981/7.

Chin, R. J., Hickman, A. H. and Thom, R. 1984, *Hyden, Western Australia, 1:250 000 geological series explanatory notes*, Sheet SI/50-04, Geological Survey of Western Australia, Record 1982/5.

Cook, G. D., Liedloff, A. C., Eager, R. W., Chen, X., Williams, R. J., O'Grady, A. P. and Hutley, L. B. 2005, 'The estimation of carbon budgets of frequently burnt tree stands in savannas of northern Australia, using allometric analysis and isotopic discrimination', *Australian Journal of Botany*, vol. 53, pp. 621–30.

Costa, P. M. 2009, Compensation for carbon stock maintenance in forests as an alternative to avoiding carbon flows, Unpublished report, Oxford Centre for Tropical Forests, Environmental Change Institute, University of Oxford, UK.

CSIRO 2007, *Australian Soil Resource Information System*, CSIRO, Canberra, <http://www.asris.csiro.au/index_ie.html>

Dean, C. and Eldridge, D. J. 2008, *Prognosis for carbon sequestration in rangelands upon destocking*, RIRDC Publication No. 08, Rural Industries Research and Development Corporation, Canberra.

Department of Environment, Water, Heritage and the Arts (DEWHA) 2005, *National Vegetation Information System. Australia—present major vegetation groups—NVIS stage 1, Version 3.0 (Albers 100m analysis product)*, Department of Environment, Water, Heritage and the Arts, Australian Government, Canberra. [Spatial data are available for download, viewed 5 February 2009, <http://www.environment.gov.au/erin/nvis/mvg/index.html> Source data for Western Australia: WA Department of Agriculture.]

Doepel, J. J. G. 1973, *Norseman, Western Australia, 1:250 000 geological series explanatory notes*, Sheet SI/51-02, Geological Survey of Western Australia, Perth.

Doepel, J. J. G. and Lowry, D. C. 1970a, *Balladonia, Western Australia, 1:250 000 geological series explanatory notes*, Sheet SI/51-03, Bureau of Mineral Resources, Australia.

Doepel, J. J. G. and Lowry, D. C. 1970b, *Zanthus, Western Australia, 1:250 000 geological series explanatory notes*, Sheet SH/51-15, Geological Survey of Western Australia, Perth.

Donohue, R. J., McVicar, T. R. and Roderick, M. L. 2009, 'Climate-related trends in Australian vegetation cover as inferred from satellite observations, 1981–2006', *Global Change Biology*, doi:10.1111/j.1365-2486.2008.01746.x.

Duncan, S., Traill, B. J. and Watson, C. 2006, *Vertebrate Fauna of the Honman Ridge – Bremer Range District, Great Western Woodlands, Western Australia*, The Wilderness Society, West Perth.

Eamus, D., McGuinness, K. and Burrows, W. 2000, *Review of allometric relationships for estimating woody biomass for Queensland, the Northern Territory and Western Australia*, NCAS Technical Report No. 5a, Australian Greenhouse Office, Canberra.

Fensham, R. J. 2005, 'Monitoring standing dead wood for carbon accounting in tropical savanna', *Australian Journal of Botany*, vol. 53, pp. 631–8.

Fensham, R. J., Fairfax, R. J., Holman, J. E. and Whitehead, P. J. 2002, 'Quantitative assessment of vegetation structural attributes from aerial photography', *International Journal of Remote Sensing*, vol. 23, pp. 2293–317.

Gallant, J. C. and Dowling, T. I. 2003, 'A multi resolution index of valley bottom flatness for mapping depositional areas', *Water Resources Research*, vol. 39, p. 1347.

Gee, R. D. 1979, *Southern Cross, Western Australia, 1:250 000 geological series explanatory notes*, Sheet SH/50-16, (First edition), Geological Survey of Western Australia, Perth, Record 1979/5.

Geological Survey of Western Australia 2007, *Atlas of 1:250 000 Geological Series Map Images, Western Australia*, April 2007 update, WA Department of Industry and Resources, Perth, <www.doir.wa.gov/gswa>

Geoscience Australia 1993, *Australian Land Tenure 1993*, National Mapping Division, Geoscience Australia, Canberra, <http://www.ga.gov.au/>

Geoscience Australia 2009, *What Causes Bushfires?*, Geoscience Australia, Canberra, viewed 3 March 2009, <http://www.ga.gov.au/hazards/bushfire/causes.jsp>

Geoscience Australia 2010, *Landsat Continental Mosaic (AGO) Product Suite*, Geoscience Australia, Canberra, <http://www.ga.gov.au/remote-sensing/get-satellite-imagery-data/ordering/pricing/landsat-continental-mosaic/>

Gifford, R. M. 2000, *Carbon contents of above-ground tissues of forest and woodland trees*, NCAS Technical Report No. 22, Australian Greenhouse Office, Canberra.

Gill, A. M. 1975, 'Fire and the Australian flora: a review', *Australian Forestry*, vol. 38, pp. 4–25.

Gower, C. F. and Bunting, J. A. 1976, *Lake Johnston, Western Australia, 1:250 000 geological series explanatory notes*, Sheet SI/51-01, Bureau of Mineral Resources and Geological Survey of Western Australia, Perth.

Grierson, P. F., Adams, M. A. and Attiwill, P. M. 1992, 'Estimates of carbon storage in the above-ground biomass of Victoria's forests', *Australian Journal of Botany*, vol. 40, pp. 631–40.

Grierson, P., Williams, K. and Adams, M. 2000, *Review of unpublished biomass-related information: Western Australia, South Australia, New South Wales and Queensland*, NCAS Technical Report No. 25, Australian Greenhouse Office, Canberra.

Groves, R. H. 1965, 'Growth of heath vegetation II. The seasonal growth of a heath on ground-water podzol at Wilson's Promontory, Victoria', *Australian Journal of Botany*, vol. 13, pp. 281–9.

Groves, R. H. and Specht, R. L. 1965, 'Growth of heath vegetation I. Annual growth curves of two heath ecosystems in Australia', *Australian Journal of Botany*, vol. 13, pp. 261–80.

Hadlington, P. 1987, *Australian Termites and Other Common Timber Pests*, UNSW Press, Kensington, NSW.

Harms, B. P., Dalal, R. C. and Cramp, A. P. 2005, 'Changes in soil carbon and soil nitrogen after tree clearing in the semi-arid rangelands of Queensland', *Australian Journal of Botany*, vol. 53, pp. 639–50.

Harrington, G. 1979, 'Estimates of above-ground biomass of trees and shrubs in a *Eucalyptus populnea* F. Muell. woodland by regression of mass on trunk diameter and plant height', *Australian Journal of Botany*, vol. 27, pp. 135–47.

Helms, R. 1892–96, 'Anthropology. Scientific results of the Elder exploring expedition', *Transactions of the Royal Society of South Australia*, vol. XVI, pp. 237–332.

Henson, P. and Blewett, R. 2006, 'Going for gold in the eastern Yilgarn', *AusGeo News*, vol. 82 (June).

Hopkins, A. J. M. and Robinson, C. J. 1981, 'Fire induced structural change in a Western Australian woodland', *Australian Journal of Ecology*, vol. 6, pp. 177–88.

Houghton, R. A. 2007, 'Balancing the global carbon budget', *Annual Review of Earth and Planetary Sciences*, vol. 35, pp. 313–47.

Hunter, W. M. 1989, *Explanatory notes on the Boorabbin 1:250 000 geological sheet, Western Australia*, (Second edition), Geological Survey of Western Australia, Perth, Record 1989(3).

Hutchinson, M. F. 2005, *ANUCLIM ver. 5.1*, Centre for Resource and Environmental Studies, The Australian National University, Canberra, <http://cres.anu.edu.au/outputs/anuclim.php>

Hutchinson, M., Stein, J., Stein, J. L., Anderson, H. and Tickle, P. 2008, Geodata 9 Second DEM and D8 User Guide, Fenner School of Environment and Society, The Australian National University and Geoscience Australia, Canberra, <http://fennerschool.anu.edu.au/publications/software/>

Ilic, J., Boland, D., McDonald, M., Downes, G. and Blakemore, P. 2000, *Woody density phase 1—state of knowledge*, NCAS Technical Report No. 18, Australian Greenhouse Office, Canberra.

Intergovernmental Panel on Climate Change (IPCC) 2007, *The Fourth Assessment Report Climate Change 2007: Synthesis report,* Intergovernmental Panel on Climate Change, Geneva, <http://www.ipcc.ch/>

Jobbágy, E. G. and Jackson, R. B. 2000, 'The vertical distribution of soil organic carbon and its relation to climate and vegetation', *Ecological Applications*, vol. 10, pp. 423–36.

Jones, R. 1968, 'Estimating productivity and apparent photosynthesis from differences in consecutive measurements of total living plant parts of an Australian heathland', *Australian Journal of Botany*, vol. 16, pp. 589–602.

Keith, H., Mackey, B., Berry, S., Lindenmayer, D. and Gibbons, P. 2010. 'Estimating carbon carrying capacity in natural forest ecosystems across heterogeneous landscapes: addressing sources of error', *Global Change Biology*, vol. 16. pp.2971-2989.

Kern, A. M. 1995, *Hydrogeology of the Boorabbin 1:250 000 sheet: Western Australian Geological Survey*, 1:250 000 Hydrogeological Series Explanatory Notes, Geological Survey of Western Australia, Perth.

Kriewaldt, M. 1967, *Kalgoorlie, Western Australia, 1:250 000 geological series explanatory notes*, Sheet SH/51-09, Bureau of Mineral Resources and Geological Survey of Western Australia, Perth.

Lawford, G., Stanley, S. and Kilgour, B. 1998, *Population Centres,* National Geoscience Dataset, Geoscience Australia, Canberra, <http://www.ga.gov.au/>

Linacre, E. and Hobbs, J. 1977, *The Australian Climatic Environment*, John Wiley & Sons, Brisbane.

Looby, M. J. 2007, *Tree hollows, tree dimensions and tree age in Eucalyptus microcarpa Maiden (grey box) in Victoria*, Masters thesis, March 2007, School of Resource Management, Faculty of Land and Food Resources, The University of Melbourne, Vic.

Low, A. B. and Lamont, B. B. 1990, 'Aerial and below-ground phytomass of Banksia scrub-heath at Eneabba, south-western Australia', *Australian Journal of Botany*, vol. 38, pp. 351–9.

Lowry, D. C. 1970, *Culver, Western Australia, 1:250 000 geological series explanatory notes*, Sheet SI/51-4, Geological Survey of Western Australia, Bureau of Mineral Resources, Geology and Geophysics, Perth.

Lowry, D. C. and Doepel, J. J. G. 1974, *Malcolm–Cape Arid, Western Australia, 1:250 000 geological series explanatory notes*, Sheet SI/51-7, Geological Survey of Western Australia, Perth.

Mackey, B. G., Keith, H., Berry, S. L. and Lindenmayer, D. B. 2008, *Green Carbon: The role of natural forests in carbon storage. Part 1. A green carbon account of Australia's south-eastern eucalypt forests, and policy implications*, ANU E Press, Canberra.

Mackey, B. G., Lindenmayer, D. B., Gill, A. M., McCarthy, A. M. and Lindesay, J. A. 2002, *Wildlife, Fire and Future Climate: A forest ecosystem analysis*, CSIRO Publishing, Melbourne.

Mitchell, K. 2007, *Quantitative Analysis by the Point-Centred Quarter Method. Version 2.15*, Self-published, <http://people.hws.edu/mitchell/PCQM.pdf>

Morgan, K. H. and Peers, R. 1973, *Esperance–Mondrain Island, Western Australia, 1:250 000 geological series explanatory notes*, Sheet SI/51-06, 10, Bureau of Mineral Resources, Australia.

National Wildfire Coordinating Group (NWCG) 2005, *Wildfire Origin and Cause Determination Handbook*, NWCG Handbook 1, National Wildfire Coordinating Group Fire Investigation Working Team, <www.nwcg.gov>

Newby, K. R., Western Australian Museum and Western Australia Biological Surveys Committee 1984, *The Biological Survey of the Eastern Goldfields of Western Australia. Volumes 1–12*, Western Australian Museum, Perth.

Ogden, J. 1978, 'On the dendrochronological potential of Australian trees', Australian Journal of Ecology, vol. 3, pp. 339–56.

Paget, M. J. and King, E. A. 2008, *MODIS land data sets for the Australian region*, CSIRO Marine and Atmospheric Research internal report No. 004, CSIRO Marine and Atmospheric Research, Canberra, [for further information with regard to this data set, contact: Matt Paget, <matt.paget@csiro.au> or Edward King, <edward.king@csiro.au>].

Raison, J., Keith, H., Barrett, D., Burrows, B. and Grierson, P. 2003, *Spatial estimates of biomass in 'mature' native vegetation*, NCAS Technical Report No. 44, Australian Greenhouse Office, Canberra.

Recher, H., Davis, W. jr, Berry, S., Mackey, B., Watson, A. and Watson, J. 2007, 'Conservation inverted: birds in the Great Western Woodlands', Wingspan, vol. 17, no. 4, pp. 16–19.

Roderick, M. L., Noble, I. R. and Cridland, S. W. 1999, 'Estimating woody and herbaceous vegetation cover from time series satellite observations', Global Ecology and Biogeography Letters, vol. 8, pp. 501–8.

Scanlan, J. C. 1991, 'Woody overstorey and herbaceous understorey biomass in *Acacia harpophylla* (brigalow) woodlands', Australian Journal of Ecology, vol. 16, pp. 521-9.

Siemon, G. R. and Kealley, I. G. 1999, *Goldfields timber research project*. Report by the Research Project Steering Committee, Department of Commerce and Trade, Goldfields Esperance Development Commission, Department of Conservation and

Land Management, Goldfields Specialty Timber Industry Group Inc., Curtin University, Kalgoorlie Campus, WA.

Skjemstad, J. O., Clarke, P., Taylor, J. A., Oades, J. M. and McClure, S. G. 1996, 'The chemistry and nature of protected carbon in soil', *Australian Journal of Soil Research*, vol. 34, pp. 251–71.

Sofoulis, J. 1966, *Widgiemooltha, Western Australia, 1:250 000 geological series explanatory notes*, Sheet SH/51-14, (First edition), Bureau of Mineral Resources and Geological Survey of WA, Perth.

Specht, R. L. 1966, 'The growth and distribution of mallee-broombush (*Eucalyptus incrassata–Melaleuca uncinata association*) and heath vegetation near Dark Island Soak, Ninety-Mile Plain, South Australia', *Australian Journal of Botany*, vol. 14, pp. 361–71.

Specht, R. L. 1981, 'Major vegetation formations in Australia', in A. Keast (ed.), *Ecological Biogeography of Australia*, Dr W. Junk, The Hague, pp. 165–297.

Specht, R. L., Rayson, P. and Jackman, M. E. 1958, 'Dark Island heath (Ninety-Mile Plain, South Australia) VI. Pyric succession: changes in composition, coverage, dry weight, and mineral nutrient status', *Australian Journal of Botany*, vol. 6, pp. 59–88.

Thom, R., Lipple, S. L. and Sanders, C. C. 1977, *Ravensthorpe, Western Australia, 1:250 000 geological series explanatory notes*, Sheet SI/51-05, Bureau of Mineral Resources and Geological Survey of Western Australia, Perth.

Tongway, D. J. and Hindley, N. L. 2004, *Landscape Function Analysis: Procedures for monitoring and assessing landscapes*, CSIRO Sustainable Ecosystems, Canberra.

United Nations Framework Convention on Climate Change (UNFCCC) 2007, Reducing emissions from deforestation in developing countries: approaches to stimulate action decision 2/CP.13, UN Framework Convention on Climate Change, Conference of the Parties 13, Bali, Indonesia.

Warde, W. and Petranka, J. W. 1981, 'A correction factor for missing point-center quarter data', *Ecology*, vol. 62, pp. 491–4.

Wellington, A. B. and Noble, I. R. 1985, 'Post-fire recruitment and mortality in a population of the mallee *Eucalyptus incrassata* in semi-arid, south-eastern Australia', Journal of Ecology, vol. 73, pp. 645–56.

Western Australia Forests Department 1934–70, *Report on the Operations of the Forests Department for the Year Ended 30th June*, Individual reports for the years 1934 to 1970, Government Printer, Perth.

Western Australia Forests Department 1969, *Fifty Years of Forestry in Western Australia*. Supplement to the 1968/69 report on the operations of the Forests Department for the year ended 30th June, Western Australia Forests Department, Perth.

Williams, I. 1970, *Kurnalpi, Western Australia, 1:250 000 geological series explanatory notes*, Sheet SH/51-10, Bureau of Mineral Resources, Australia.

Williams, R. J., Zerihun, A., Montagu, K. D., Hoffman, M., Hutley, L. B. and Chen, X. 2005, 'Allometry for estimating aboveground tree biomass in tropical and subtropical woodlands: towards general predictive equations', *Australian Journal of Botany*, vol. 53, pp. 607–19.

Woldendorp, G. and Keenan, R. J. 2005, 'Coarse woody debris in Australian forest ecosystems: a review', *Austral Ecology,* vol. 30, pp. 834–43.

Wynn, J. G., Bird, M. I., Vellen, L., Grand-Clement, E., Carter, J. and Berry, S. L. 2006, 'Continental-scale measurement of the soil organic carbon pool with climatic, edaphic, and biotic controls', *Global Biogeochemical Cycles*, 20, GB1007, doi:10.1029/2005GB002576.

Zerihun, A., Montagu, K. D., Hoffmann, M. B. and Bray, S. G. 2006, 'Patterns of below- and aboveground biomass in *Eucalyptus populnea* woodland communities of northeast Australia along a rainfall gradient', *Ecosystems*, vol. 9, pp. 501–15.

www.ingramcontent.com/pod-product-compliance
Lightning Source LLC
Chambersburg PA
CBHW051308270326
41928CB00027B/3452